ADVENTURES IN CLIMATE SCIENCE

Woodslane Press Pty Ltd
10 Apollo Street
Warriewood, NSW 2102
Email: info@woodslane.com.au
Tel: 02 8445 2300 Website: www.woodslanepress.com.au

First published in Australia in 2023 by Woodslane Press
© 2023 Woodslane Press, text and internal images © 2023 Wendy Bruere
and contributors. Cover images: Quinn Zeda/Shutterstock.com,
andreiuc88/Shutterstock.com, Ethan Daniels/Shutterstock.com,
Constellaurum/Shutterstock.com, PrasongTakham/Shutterstock.com.

The information in this publication is based upon the current state of commercial and industry practice and the general circumstances as at the date
of publication. Every effort has been made to obtain permissions relating to
information reproduced in this publication. The publisher makes no repre-
sentations as to the accuracy, reliability or completeness of the information
contained in this publication. To the extent permitted by law, the publisher
excludes all conditions, warranties and other obligations in relation to the
supply of this publication and otherwise limits its liability to the recommended retail price. In no circumstances will the publisher be liable to any
third party for any consequential loss or damage suffered by any person
resulting in any way from the use or reliance on this publication or any
part of it. Any opinions and advice contained in the publication are offered
solely in pursuance of the author's and publisher's intention to provide
information and have not been specifically sought.

Book design by Mike Ellott

ADVENTURES IN CLIMATE SCIENCE

SCIENTISTS' TALES FROM THE FRONTIERS OF CLIMATE CHANGE

EDITED BY **WENDY BRUERE**

WOODSLANE
PRESS

CONTENTS

FOREWORD

BY

DR KARL KRUSZELNICKI

The study of Climate was not invented by the 'greenies' specifically to annoy Rupert Murdoch.

The Science probably began with the fabulously wealthy Medici family in 1645. They measured temperatures (with calibrated thermometers) across Europe. They knew that the temperature had financial repercussions (crop failures, etc.) and they were in for the long haul.

Surprisingly, in the 1970s and 1980s, the best research into Climate Science was being done by Big Fossil Fuel!

Back in 1973, 'Global Warming' was obvious to the world's largest re-insurance company, Munich Re. The payouts from extreme weather events were getting more costly. Munich Re simply increased the premiums they charged, so they wouldn't make a loss. (Nothing personal, it was strictly business.)

By 1982, Big Fossil Fuel had predicted that, by the year 2020, under the scenario of Business-As-Usual, the carbon dioxide levels would rise to 415 parts per million (ppm) and the temperature would rise by 0.9 degrees Celsius—and they were astonishingly close.

In 1990, there were two big events in Climate Science.

First, climatologists had enough data to be very confident that we humans were responsible for the increasing temperatures they were measuring.

Second, Big Fossil Fuel chucked a U-turn and started spending up to US$1 billion each year to now deny the science of Climate Change.

The finances are 'interesting'.

Each year, out of a total world Gross Domestic Product of about US$85 trillion, Big Fossil Fuel gets a 'subsidy' (i.e. free money that does not have to be paid back) of US$5.9 trillion. (Yes, that is on top of their colossal profits.) That works out to about seven percent of all the money generated on Earth.

Amazingly, to stop and reverse Global Warming, and bring carbon dioxide and the climate back to 20th Century levels, would take only two percent of the total world GDP. Humanity would benefit. And we could even go further and spend the remaining five percent of the total world GDP on education, health, welfare and other worthy projects.

Today, we humans generate about 20 TeraWatts of power, but the Sun dumps much more than that onto the Earth—40,000 TeraWatts. Greenhouse Gases changed the balance of that torrent of power from the Sun by a tiny amount—and bingo, Climate Change.

But the Message of Good Hope is that we know that we can reverse Climate Change with today's technologies.

Greenhouse gases (overwhelmingly from fossil fuels) trap about 600,000 Hiroshima-sized atom bombs' worth of heat each day. You can get away with that only for a little while. But now, after several decades, this trapped heat has warmed both the lower atmosphere and the upper oceans by about one degree Celsius.

This collection of stories comes from the next generation of scientists working at the frontiers of Climate Change. Their work covers an amazing range—from bushfires to oceanography, tropical forest ecology to geophysics, endangered species to glaciology and more.

Read their stories and be empowered.

DR KARL KRUSZELNICKI 2023

FIRES OF THE FUTURE

―――― BY ――――

PHILIP ZYLSTRA

Dr Philip Zylstra *is a fire scientist and forest ecologist who looks at how plant species interact with weather and the terrain to affect fire behaviour, and how this in turn affects the survival of flora and fauna. His work has led to groundbreaking advances in modelling fire behaviour and landscape analysis. Phil believes that the questions we most need answered are often the ones that challenge paradigms and policies, but that the scale of our impact on the earth means we no longer have the luxury of avoiding controversy. His current focus is on understanding how fire management can be adapted to work with the 'ecological controls' that have enabled forests to persist over evolutionary timescales.*

One of the things about being lowered on a long cable under a helicopter is that the downdraft can start you spinning, which makes you dizzy and can play havoc with the stability of the aircraft. If that goes badly enough as they're lowering you through the gap in a tall forest canopy somewhere, you could theoretically swing into a treetop which could theoretically get the helicopter tied to the tree by your cable, in which case (theoretically), they need to cut the cable. That never happened to me or any other Remote Area Fire Trained (RAFT) crew I had heard of. To my knowledge, the NSW National Parks and Wildlife Service have not lost a RAFT crew member since they pioneered the approach in Australia; they're just very careful and train their crews well.

When people think of firefighting, they often picture people standing by a truck, armed with nothing more than a hose and with every fibre of their being shouting, 'You shall not pass!' to the flames that tower over them and blacken the bright day with the fumes of hell. That wasn't us, we didn't necessarily have hoses. RAFT crews get sent into country that you can't reach from the road, which means that we only have what we can carry with us—and we're almost always carrying it through the steepest country there is. For that reason, they don't send us in on the wild days; instead, RAFT go in straight after the storm passes and put out the lightning strike or spend weeks wandering the fire edge that looks dead. We feel the hot ground for hotter patches, touch the base of every tree or log that looks very much like it could be smouldering inside, then break it open and scrape out the coals and—if one is available—we call in a chopper with a 'Bambi Bucket' to make sure the fire is dead.

There is a particular perspective you gain after enough time on foot with an experienced crew, watching what a fire does in different conditions or reading its behaviour from burn patterns with the eyes of someone who is in the wild with it (with no quick escape). It becomes important to pay attention. I had paid attention to fire from the time I first started working with it. Those were days I spent alone in the high Monaro grasslands, burning snowgrass for the graziers in late winter to give them fresh growth for stock. To begin with, I walked from tussock-to-tussock lighting each with the drip torch. We didn't want a running fire in that country; all these old men had seen one some years before and remembered it too well. One had been trapped and didn't like to talk about it. The speed it could cross the grasslands with a gale behind it was terrifying, as were the burning pellets of sheep turds that flew in the wind across roads and firebreaks. I kept it small. I had a lot of country they wanted to be burned though, and that meant that I started to look for the exact situation where fire would spread from one tussock to the next, and where it wouldn't; conditions that would let me burn very small runs. That particular insight shaped the way I look at fire.

An important and weirdly controversial part of that insight was that what changed flames from little things into big things was all about the plants. There is a big difference between a flame from one burning tussock and a flame from a patch of tussocks burning together. Or, for that matter, a flame from a burning tree. Closely packed tussocks give much bigger, faster flames.

Burning trees give much, much larger flames. Crown fires release so much energy that if the atmospheric conditions are right, they can create storms with lightning, rain, and even tornadoes. The question though is whether they will burn—can flame make it up there? To get from the ground into a tree, small flames have to light bigger things like shrubs and climb their way up, so a crown fire can only happen when there is a ladder of plants beneath it. If it doesn't have that, then the tree not only doesn't burn, it actually slows the wind beneath it and calms the flames. Without wind, one burning tussock doesn't set its neighbour on fire, because to do that the flame needs wind or slope to tilt it over toward its neighbour. So, plants close to the ground make fires bigger, plants high above the ground make fires smaller.

This is where it gets controversial. Back in the 60s, a CSIRO fella by the name of Alan McArthur suggested that what makes fire more or less severe is the weight of dead leaves, bark and twigs on the ground—something he called the fuel load. He was quite clear at the time that this was a first guess and may be proved 'drastically wrong' with some more investigation, but it was a starting point. It turns out it was also a stopping point. As is the way of things, an industry was built around this: policies were written, governments developed Key Performance Indicators and talkback radio hosts harnessed the power of angry old men. We know what to blame for the houses and lives lost to bushfires: the bush. Unless we step in and burn it, the bush will keep accumulating that fuel load until it is like a bomb waiting to go off. It needs a firm hand to save it from itself, it needs hardened men who are not afraid to do what must be done. The problem was that even when they burnt, the bushfires kept coming. Every time the answer was the same: more burning was needed.

Working in fire management, I had to conduct a lot of these burns, then I had to go back later and fight the fires that burnt through the country we had burnt to stop them. What bothered me when I did that was that the places where we had burnt now had a nice ladder of plants because fire does something to the bush. Yes, it burns away the leaf litter and that's sparse for a couple of years, but it also germinates the shrubs, and those shrubs put a lot of dense, foliage down close to the ground where it can easily ignite and turn little flames into big ones. We had reduced the 'fuel load' for a couple of years, but we had made fires worse for decades. However, if those forests are left 'unmanaged' for decades, the shrubs thin again and the forests create the safe environment that they need for long-term survival. It turns out that the forests didn't need us to save them from themselves, they needed us to sit down and learn.

I didn't realise this early on. The day I sat in the shearing shed at smoko and told the others that I was going to Uni to study science, all I knew was that I was curious about the way the world worked and I wanted to know more. I certainly didn't want to work for National Parks because, as everyone knew, they were responsible for every fire, weed and dead sheep there was. And yet, less than a decade later, in 2003, I was there in the National Parks office as more than 100 fires raged across the Snowy Mountains and Victorian Alps, all lit by a single storm front that passed through without a drop of rain. When the smoke finally settled and my RAFT crew patrolled the remote fire edges looking for smouldering remains that might start a new fire, 1.5 million hectares of the Alps had burned.

Fires of that size had happened more than 60 years before, but something critical was different this time. In the old days, the hundred ignitions all came from fires that people had lit; this time they came from lightning. There have always been lightning fires in the Snowies, but there are rarely more than a handful of ignitions at once because the storms are smaller and bring rain, and the landscape has natural controls in wet gullies and country that doesn't burn well. This year, ignitions bathed the mountains and immediately became monsters because the storms held no rain. I remember the day well, the eerie haze in the air. Seasoned fire-tower observers reported lightning from a blue sky. This was something broken in the climate, but it wasn't just the storms.

The cold air over Antarctica circles the pole in a vortex, confining strong, dry winds to the far south. In 2002, air from our warming atmosphere gathered over the pole, collapsing the vortex completely for the first time that we know of. Dry air pushed northward, heating up as it passed over the deserts, baking the forests, changing snow country into a drought-dry landscape ready for a spark. This is the thing with climate change: we watch the temperature climb each decade in fractions of a degree and it feels like change is incremental. We're wrong. Our changing climate oscillates around currents and massive cycles, but every now and then, everything bad comes together at once and the world we know falls to pieces.

That happened again in 2019. Again, hot air gathered over Antarctica, and again the vortex began to fall apart. This time, the effects dwarfed those of 2003.

There are places in the tropics and sub-tropics of Australia where the same bird calls sound that were heard by the dinosaurs. The same trees grow and flowers bloom that once covered the magnificent land of Gondwana, of which Australia was a small part. Fire had no part in that world, except for the volcanoes like the one which formed the massive caldera behind Tweed Heads on the Queensland border. It may be that no fire has entered those forests since the lava dried and the forests of Nightcap and the Border Ranges grew. Islands of ancient Gondwana holding out against the drying continent. That changed in 2019. The night cap of cloud stopped covering the plateau because the air was dry. So dry that fire was able to spread there and kill trees like those in the only living stand of Nightcap Oak on earth.

Dry Antarctic air could not reach tropical Queensland, but something else did. The incidence of severe cyclones has more than doubled since Cyclone Tracy levelled Darwin. That was a Category 4, but Cat 5 cyclones are so frequent now that Tracy would seem small. The tropical air is charged with heat and moisture, and that means power. Enough power to level the ancient Gondwanan remnants of Kutini-Payamu, the Iron Range in north Queensland. Those trees had maintained a wet microclimate for aeons, but when that was broken, so was the power of the forest to withstand fire. It dried just as the southern forests dried and when lightning arrived even Kutini-Payamu burned. All the bad things at once, all compounding each other.

We're not powerless though, not if we're prepared to learn. Fire burned those Gondwanan remnants, but it trickled through that ancient forest with small flames. Flames small enough that a RAFT crew like the ones I worked in could contain them with rake hoes, and without the need for huge back-burns or massive planes. We didn't do that, but we could. We need to learn to understand our forests. We need to know where the old stands still remain and to nurture others to restore their range. These places where the forests have rebuilt from devastation are our hope; when fire comes—as it certainly will—we can cooperate with and reinforce the controls that forests have exerted over fire since the shape of the world was changed. Our changing climate is a tsunami of a magnitude that we may not yet comprehend, but there are other forces in our ancient land. If we hold onto the old colonial thinking that we need to bash the bush into submission, then we have sided with the changing climate to break the forests' defenses. Now is the time to come of age and choose a side.

OXYGEN THIEVES

BY

ANDREW MEIJERS

Raised in Hobart, Australia, **Dr Andrew Meijers** first became involved with physical oceanography and the Southern Ocean in his honours year at the University of Tasmania in 2004. His work there with high-resolution models led him to his PhD where he combined satellite observations with in-situ observations to infer changes in the heat content of the Southern Ocean interior. After a few short postdocs at the Australian CSIRO Marine and Atmospheric centre, Andrew was offered a post at the British Antarctic Survey in Cambridge, UK. Since being at BAS Andrew has participated in and led numerous voyages in the Southern Ocean and broadened his science focus to encompass the modelling and observations of Southern Ocean circulation and dynamics from the ice shelves to the subtropics. He has taken on more and wider leadership roles, culminating in leading large multi-disciplinary research programmes incorporating numerous people and institutes from across the UK and Europe. These programmes have strong climate focuses, looking at the surprisingly central role that the Southern Ocean and Antarctic ice sheet plays in our global climate, and its potential changes in the coming decades and centuries.

T he Royal Research Ship *James Clark Ross* bobs amid the shadows of snow-capped peaks leaning in on all sides. Her slab-sided red hull stands in stark contrast to the greys, blues and greens of the overcast South Georgian fjord, and her white superstructure, now somewhat shabby after a hard season south, reflects the tumbled glaciers flowing down into the bay.

I stand on the bridge with my binoculars trained on the ice wall at the head of the bay, or more accurately, the tiny black dots in front of the sheer ice cliff. Magnified, the dots resolve into two small RIBs, rigid inflatable boats moving slowly across the lumpy waters in opposite directions along the glacial face. Occasional white caps whip the tops off waves and show that the storm raging outside this bay can be felt despite the shelter provided by mountains up to 3000 metres high. Ralph, the captain, stands next to me in his officer's whites with radio in hand, also staring through binoculars at the ice cliff. Ralph is an imposing figure, large in all ways, despite the hunch that many a tall man instinctively develops aboard working ships. After his size, the first thing people notice is his meticulously manicured moustache that extends several inches on either side, curled and waxed to movie villain points.

'Make sure you're keeping clear of the face, 100 meters please,' he intones again.

'Aye, aye,' crackles back over the sound of revving motors and a water-smacked hull.

I can hear the terseness in the reply of the Chief Officer; this is the third time Ralph has reminded them to keep back. His worry is that the RIBs will stray too close and one of the regular calvings of the glacial face will send thousands of tonnes of ice crashing down on them. Carola, the tough German Chief, is more than aware of the risk. She is also very used to handling scientists and steering us clear of the hazards our sometimes-unwitting sampling requests might incur. And that is what we are doing here, sampling. In each of the small boats are two scientists, frantically filling up sample vials of seawater from directly, or as directly as the careful Carola will allow, where the glacial meltwater merges with the waters of the great Southern Ocean beyond South Georgia. This has to be done in small boats, as the ship itself cannot get close enough to the glacial front and would stir up the water column in any event, ruining any measurements of the thin layer of ice-freshened melt water sitting above the saltier seawater.

This is JR16005, the fifth voyage of the British Antarctic Survey's RRS *James Clark Ross,* or JCR to all who sail and love her, of the 2015/16 Antarctic Season. It is also the last voyage of this season and we are on our final few science days before heading back to the Falkland Islands and leaving the JCR and her crew to the long trek back up the Atlantic. These science days consist of round-the-clock sampling and are the bread and butter of what we physical oceanographers do. They are the gold standard by which all our other tools; satellites, robotic floats, drifters, moorings and buoys are calibrated, and the only way in which we get physical water samples to undertake more complex chemical analysis on.

The previous day, I had made the call to run from the storm now pounding the ocean around South Georgia. Instead of sitting miserably with the ship hove to, head to the wind and pitching through endless lines of rollers, I thought that we would instead use our precious remaining science days sampling the sheltered waters among the fjords of the huge sub-Antarctic island. As Principal Scientific Officer, or chief scientist, I have to balance the delivery of our science goals with the conditions, and work with the ship's captain to make it achievable and safe. So here we are, using our time in the best way we can, sampling the surface salinity and water oxygen isotopes in front of glaciers.

'Heading back.'

The radio jerks us back to the here and now and we see first one RIB and then the other reach the end of their line and then turn back towards the ship.

Faces glowing red from the cold and the exertion of struggling from heavy boat suits greet me as I head downstairs into the ship's wet lab. Clinking racks of stoppered bottles are passed from hand to hand and disappear to labs and holds for later analysis. Soggy log sheets are carefully spread over the metal sample tables before they disintegrate.

'Check this out, Skip!'

I'd like to think it's short for 'Skipper', but I'm pretty sure Ollie is just poking gentle fun at my Australian accent. Grinning through a piratical cruise beard, he hands me his GPS mapper, still crusted with dried sea salt. The ruggedised screen looks like an ancient Gameboy but squinting at the green and black lines I can make out the mapped outline of the glacier front and the small icons showing sample locations. Most of the sample marks are well inside the glacier outline, where a wall of ice should be. Sadly, this glacier, the Neumayer, has retreated over five kilometres since the 1970s and the maps on the GPS unit haven't been able to keep up.

This is one of the reasons we are here. Well, a symptom of the wider reason we are here, at least. This voyage has two goals. The first is to reoccupy the A23 hydrographic section, a line that traces its way south from South Georgia and goes as far into the Weddell Sea as possible until the sea ice stops us. 'A' is for 'Atlantic' and A23 is one of the many lines drawn by scientists back in the 80s and 90s as part of the great World Ocean Circulation Experiment; the first really rigorous attempt to sample the ocean that covers most of our planet. Many lines were only visited once, but others like this receive ship expeditions every few years and provide a critically important record of how the ocean is changing. A23 crosses the southern side of the great Antarctic Circumpolar Current that rings Antarctica and passes over the subsurface mountain chain that separates the Scotia Sea from the Weddell Sea proper. This selection of dynamically distinct regions contained within a fairly short distance is what keeps us coming back.

The other reason is that these painstakingly gathered lines of hydrographic data show that the ocean around Antarctica is changing. The surface layers of the ocean here are getting fresher, and fresher water is more buoyant, altering the ocean circulation as a consequence. Although it feels like the motion of the Southern Ocean has little bearing on our lives, this could not be further from the truth. The ocean stores vastly more energy than the atmosphere and has absorbed more than 90 percent of all the extra heat that humans have created since the start of the industrial revolution. The bit we hear spoken about so much, on the news, in the papers and online—the atmospheric warming that the Paris Accord strives to hold to 1.5 degrees—represents only a few percent of this extra heat. A few more percent warm the ice caps and glaciers, but all the rest is in the ocean. The ocean also sponges up about 30 percent of the extra CO_2 that humans have been emitting. This drives ocean acidification but shields the atmosphere from an even stronger greenhouse effect. So how does it all get there? The answer is mostly via the Southern Ocean. In most parts of the world, the ocean's deep waters are trapped below the warm surface layers, with little exchange between them. In the Southern Ocean, however, the cold temperatures, sea-ice formation, strong winds and the only open pathway where the ocean can flow right around the Earth, combine to pull deep waters to the surface, as well as sink newly cooled waters deep into the abyss. This constantly renews the ocean, and permits the introduction of anthropogenic, human-induced, changes into the deepest layers of the ocean.

The question we are striving to answer now is: Will this overturning and heat and carbon uptake continue? Will the Southern Ocean continue to sweep our climate guilt under the rug? Or will it one day give up and stratify like the rest of the ocean, accelerating atmospheric warming? We know the winds are changing over the Southern Ocean, due to a combination of ozone depletion and greenhouse warming, so how will this alter the vital overturning and renewal of the ocean? How will the surface freshening we see impact this and where is that freshening coming from anyway? And that is the second reason we are here: to investigate the source of the Southern Ocean freshening.

'PSO to the captain's office please. PSO to the captain.' The ship's Tannoy snaps me out of oceanographic exposition.

'Great stuff, definitely a figure for the cruise report,' I say, handing the GPS back to Ollie. I shrug as the scientists' raised eyebrows around me query the call to the old man's office.

I leave them with 'Dunno. Maybe we've run aground?' as I duck through the door, heading for the six flights of stairs from the deck to the captain's dayroom. Information is always in short supply on a large working ship and the Principal Scientific Officer being called to talk with the captain will start the inevitable rumour mill tearing through the crew and scientists about what change of plans may be brewing.

Puffing a little from the combination of stairs and seven weeks of excellent ship meals, I knock on Ralph's office door and enter. It's less of an office and more a small living room with a desk in it. Comfortable couches and bookshelves occupy the rest of the space, which is opulent in comparison to the regular scientist's cabins, where two or three people live in a single narrow room. I sit down at the polished desk across from Ralph. It is fastidiously empty, except for a prominently placed moustache grooming kit.

'Ah, Andrew. Good. Jumping right in,' he begins, 'I've had a call from the FCO…'

'The what?' I'm continuously being surprised by new nautical terms, but this is one I probably should have already known.

'Sorry, the Foreign and Commonwealth Office… you know, our other bosses?'

I'm dimly aware that the British Antarctic Survey belongs to two government departments, one which is concerned with sciences and one, the FCO, with flag-waving, maintaining the British presence in the South Atlantic and supporting the UK Antarctic claim.

'Um, sure. Yes, right. What do they want with us then?'

'They've had a request from the South Georgian Government for us to investigate a suspected pirate fishing vessel. The Pharos is over in the Falklands and so we're the nearest ship. We're to jump to it.'

This is new. I've worked in and studied the Southern Ocean for around 15 years and, prior to that, growing up in Tasmania made me acutely aware of the problems of pirate fishing boats in the Southern Ocean. Often flagged by nation states operating outside the internationally agreed fishing quota system, they sometimes stray into the richer fishing grounds and plunder the nation's territorial waters. They mostly target the slow-growing and valuable stocks of Patagonian toothfish, causing significant ecological as well as economic damage. My mind immediately flicks to an incident back in 2001 when Australian fisheries undertook an incredibly low-speed pursuit of such pirates across the entire width of the Indian Ocean, ending with a combined South African-Australian military boarding action and a sunken ship. Fear for our remaining science days, and even our flights home after weeks at sea, quickly rushes in.

'What do they want us to do about it?' I blurt.

'No idea. We're to take on their fisheries officers at KEP and go and look,' Ralph says, evidently unhappy with the vague directive. 'I'm afraid science is done for today.'

The storm is still raging over and outside our little bay a few hours later as we pull around to the BAS base at King Edward Point. We are met by a small boat that transfers two stern-faced

South Georgian government officials aboard. They have to be stern-faced, as the Government of South Georgia and the South Sandwich Islands, to use their full title, is a government without a people and exists only to exercise its own administration. No-one actually lives on the islands permanently, so as its sole purpose for existence, the government takes bureaucracy to Kafkaesque levels of seriousness. And there is nothing they take more seriously than fisheries. Toothfish and squid licenses bring in many millions of pounds a year, and the government celebrates this reason for existence to such a level that they have even given themselves a holiday in its honour: 'Toothfish day', September 4, is a bank holiday. Despite there being no banks in South Georgia. Or a population to take the day off. Seriously, look it up.

As the ship turns and heads north in the rapidly fading light the two officers brief the JCR's officers, clustered around a map table on the bridge. A legal fishing vessel, *Rambler*, reported a radar contact approximately 40 nautical miles north-northeast of our location. Apparently, the radar contact was moving in a slow manner consistent with trawling operations, was operating without the Automatic Identification System transponder system required by international law and refused to respond to radio hails.

'We take this report very seriously,' said the fisheries officer who was obviously taking things very seriously.

'Very seriously indeed,' said the second, who was presently taking it very seriously, but in the morning would probably have to be the island's postmaster general (and take that very seriously in turn).

'What do we do if we find them?' I ask, dreading the possibility of a long chase across the South Atlantic and the total loss of our remaining science days. 'Shall we send Wave across with a teaspoon to sort them out?' I ask, in a floundering attempt to lighten the mood. Wave is our third officer, ex-Special Boat Service, and probably knows a thing or two about boarding actions. My attempt fails to dent the granite visage of the first officer.

'Our first job is to establish contact and verify their identity, from there we will see what develops.'

Convinced by the thorough nature of the officer's proposal, we turn our attention to the fading light and mounting seas. It is fully dark as we finally leave the shelter of Cumberland Bay and into the ocean proper. The Furious Fifties are waiting for us and immediately slam the ship on its beam, sending us wallowing sideways up and down the unseen waves. Blasts of snow form horizontal white lines streaming through the searchlights that now barely even illuminate the bow through the blizzard. The blinding snow in the beams is all we can see, except when a wave face looms up out of the darkness. These waves, black with foam dripping from their tops, send the bow corkscrewing through the troughs or smash boiling green water over the focsle, spraying the bridge. As we lurch down the face of one monstrous black shape, the props break the surface, sending shudders through the entire ship. Ralph quietly relieves the fourth officer from the con and takes over himself.

The ship immediately steadies, as the captain expertly picks a course that deadens the worst of the rolling. I walk to the back of the darkened bridge to glance at the met instruments. The wind is gusting to over 60 knots, more than 100 kilometres per hour. Back standing behind Ralph, I see that he has called all four of the JCR's engines online. Normally we only run on two;

the nearly seven megawatts of power that four can generate is reserved for breaking ice. The ship is also now forcing through the waves at better than 14 knots. This is flat out for the dear old JCR: we are ordinarily restricted to a sedate 10 knots to conserve fuel. It's hard to tell in the dark, but there definitely seems to be a gleam in the eye of the captain, joy at being let off the leash in some of the worst weather the Southern Ocean can conjure.

The hours pass in silent, building tension, the frenzied howling outside broken only by snatches of murmured conversations. The bridge is pitch black to conserve our night vision, but I can feel the silent crowd growing around me as excited scientists and off-watch crew surreptitiously mill about. Occasionally I see faces floating in the dark, lit from below by the ghostly green-yellow of the large radar displays. One is Carola, frowning intently at the screen as she fiddles with the gain to pick out features from the mess of wave crests.

'Ah,' she smiles. 'There's something there.' The darkness shuffles in, craning to see the screen where she is pointing. A grainy blob, surrounded by other blobs. But as we watch it is the only one that holds steady each time the glowing arm of the radar sweeps by. I look at the range rings. The pirates are about 20 miles out and exactly where *Rambler* reported it. A frisson of excitement passes through the room. Ralph adjusts the course slightly to lay the contact dead ahead.

As the miles slowly tick down, I drift into a narratively convenient contemplation of what we may lose if this drama goes on. Tomorrow we are due to undertake long looked forward to fieldwork on South Georgia itself. Here we hope to collect our final set of water isotopes from different sources of fresh water to the Southern Ocean. These water isotopes or, more properly, the ratio of oxygen 16 to oxygen 18 isotopes bound up in the H_2O molecule of the seawater, represent a relatively new approach to finding out the sources of the freshening we are observing. The water transported in the atmosphere from the equator towards the pole rains out the relatively heavier oxygen 18 earlier in its journey south, leaving more water containing oxygen 16 to fall as snow on the Antarctic ice sheet.

This ice eventually finds its way, over thousands of years, into the ocean as meltwater and icebergs. Sea ice takes the oxygen isotope ratio from the seawater, but shifts its salinity dramatically as it freezes and melts. The deep ocean also has a unique isotopic signature, which is relatively uniform around the globe. Therefore by measuring how salty any given parcel of water is and its isotopic signature, we can estimate the contribution of the potential candidates…that is, melt from the sea-ice, glacial ice, rain/snow and the deep ocean itself. To do this relatively simple simultaneous equation, balancing the contributions from each freshwater source, we need end members; the isotopic signatures of the sources of freshwater themselves. If these relative contributions change over time, we then have candidates for the drivers of ocean freshening, and maybe further evidence for an enhanced rate of melt of the Antarctic ice sheet itself.

This voyage's second objective has been to gather these end members. We already have glacial samples from islands further south, sea ice samples from deep within the Weddell Sea and isotopes collected this morning in the embayments in front of the South Georgian glaciers. What we would like is some of this South Georgian ice itself. The possibility of a latitudinal gradient in glacial isotope concentration is something we need to consider if our calculations are to be accurate. It would also be a personal blow to miss this opportunity. South Georgia is possibly the most dramatically beautiful place I've ever seen. Ever since first setting eyes on it I've dreamed of climbing the glacial peaks above the seal and penguin infested beaches and green mossy foothills.

'Two miles,' intones Carola.

I can see the worry in her eyes. Below her pragmatic exterior lurks a true nautical humanitarian. The crews of these illegal fishers, mostly from poor nations, rarely have much choice in the matter. They are employed by distant corporations hidden behind flags of convenience and other layers of administrative obfuscation. Carola's virtues would one day see her arrested in Italy as the captain of a refugee rescue vessel, when she chose the wellbeing of her charges over the niceties of international law. Further on I see Wave, looking rather more relaxed and eager. His barroom stories of helicoptering onto drug running ships and 'rescuing' unwilling reporters from war-zones are beginning to feel a lot nearer to home right now.

'One mile,' says Carola. Ralph starts easing back the engines. 'It's huge,' a note of uncertainty enters her voice. 'We should be on top of it any second now.'

A dozen sets of eyes stare into darkness beyond, straining for anything in the spotlights, still failing to pierce the blinding snow tearing madly past. A lighter patch appears in a brief gap in the storm, everyone leans forward. Knuckles grow white around the window railings as we squint through a renewed flurry.

'There!' points Dan(i), one of our ocean computer modellers who has run away to sea. The wind drops and the snow clears away from the object, looming from the gloom.

'It's bloody massive,' says another voice in the darkness.

'It's... it's...'

'It's a fooking iceberg!' shouts a northern voice from the back, one of the engineers who has snuck up from their below decks haunt, and the mischievous glee in his voice is evident.

The room immediately deflates, every shoulder sags and I let out a puff, I'd been holding my breath without realising it. Relieved, but somehow disappointed all the same.

The tension is gone, sudden laughs and grumbling voices fill the bridge as people glance at their watches. It has gone midnight and the crowd dwindles rapidly as the sightseers drift off below to bed. Ralph makes the graceful gesture of checking behind the burg at the sheepish request of one of the fisheries officers and then heads to bed himself. I follow him down the stairs towards my cabin, exhausted from a long and tense day. Suddenly the walls tilt madly and I cling onto the railing. The ship is coming around southward, running for the shelter of South Georgia once again, and soon the wallowing and pitching resumes at the artless hand of a junior officer.

A sleepless night spent sliding up and down my bunk gives way to a grey morning in the blessedly calm waters of Cumberland Bay as the JCR moves in to dock at the base. As I watch the two fisheries officers disembark, doing their best to remain dignified, I open up my laptop to check the day's email. The first is from the South Georgia Government. Looking forward to their glowing praise I open it, but quickly turn the air blue with recently acquired nautical invective.

They've refused my glacial sampling request due to a typo in the form!

POLE TO POLE:
A JOURNEY INSPIRED– AND ALMOST ENDED– BY CLIMATE CHANGE

——— BY ———

JAMES HOOPER

Dr James Hooper summited Everest at 19 before completing a world-first expedition from the Arctic to the Antarctic using only human and natural power in 2007-2008. He was awarded the National Geographic Adventurer of the Year Award in 2008 and also became a Fellow of the Royal Geographic Society in London. He has a BSc in Geography from Kyung Hee University in Seoul and a PhD in Earth & Environmental Science from The University of Wollongong, Australia. He has worked as an Assistant Professor in the Department of Biological & Environmental Science at Dongguk University in Korea, teaching Atmospheric Science. He is currently working with multi-billion-dollar companies and property portfolios to help them reduce their environmental impacts. James is a founder and trustee of the charity, One Mile Closer. He has appeared on numerous Korean TV shows, including JTBC's Non-Summit and MBCevery1's Welcome to Korea.

We had been skiing for a couple of days from the Geomagnetic North Pole towards the major landmass of northern Greenland when we came across the first of the leads in the sea ice—a narrow channel where the ice had cracked apart. This lead was about 15 metres across, with gnarled, spiky edges.

It was early April 2007 and Rob and I were 19. We had planned this trip, the *180 Degrees Pole to Pole Expedition*, to try to become the first people ever to travel from Pole to Pole—the Geomagnetic North Pole to the Magnetic South Pole—without using any engines. We had known it was possible that we would have to navigate obstacles like this lead on the expedition, but we hadn't expected to meet one so early in the season and so far north.

The Greenland icecap rose out of the sea ice in front of us, probably 10 or 20 kilometres away. We could see the giant granite buttresses of the various headlands and, between them, the slow rise of a glacier as it ascended to around 1,000 metres. We had passed the spring equinox and the sun was in the sky 24 hours a day. It was always light, but a very stark, clear light, cold and bright, not like the lovely golden hour you get in the evenings in the mid and low latitudes. It was incredibly cold too, between -20 and -40 degrees Celsius. We needed highly specialised bulky gear to stay warm and wore big jackets with fur-lined hoods as we pulled our pulks, each weighing around 100 kilograms, across the ice.

Through the trip, we aimed to raise awareness of climate change, particularly among young people. In 2006, when we started planning, climate change was increasingly becoming recognised as a major issue by the general public, regularly spoken about in the press, and there was a growing realisation that it must be taken seriously. Rob and I wanted to find a way to take this quite heavy and nebulous scientific topic and make it accessible to younger people.

I won't lie and say our main goal wasn't to have an incredible adventure, but we were also focused on creating stories and content for people to engage with the changes that were happening around the world. Changes we were guaranteed to see as we traversed every environment on earth. We had an excellent interactive website with live content and related educational material that we updated by satellite, assisted by a small team in the UK. Schools in the UK—where we were both from—and across the world used the website for geography and biology classes.

Rob and I had met at the beginning of high school, aged 11, and over the following years discovered a mutual love of challenge and adventures. This started small, getting up before dawn to do a pre-school triathlon consisting of swimming the length of a local lake while trying to avoid the swans, then cycling a loop through the local lanes before finishing by running laps around the outside of our school. These challenges were infectious and the draw of doing something different, something novel, kept driving us to find newer, bigger challenges.

Then, at the age of 16, we saw an article about the 50th anniversary of the first ascent of Mount Everest. We knew right then this was the adventure that would define our lives for the remaining years of high school.

The next three years were a whirlwind of part-time jobs to pay for new equipment and training expeditions to the Alps and the Himalayas, all the while desperately trying to find sponsorship for the ultimate goal of Everest. Our grades suffered, but we gained a different understanding and appreciation of the world and its sensitive environments through visiting them. Neither of us received top grades in our favourite subject, Geography, but we had hiked on glaciers and seen avalanches first-hand; we had cycled across flood plains and through the scarred remains of war zones. These experiences were instrumental in developing our interest in the natural world and an appreciation of the threat posed by climate change.

In May 2006, we successfully summited Mount Everest, becoming, at the time, the youngest non-sherpas to do so. But the desire for a new adventure remained and what could be more exciting than an expedition from Pole to Pole, travelling through every type of environment on earth, all by human and natural power—ski, dogsled, sailboat and bicycle.

So, there we were, just a couple of days into the expedition, facing the gash of inky black water cutting through the almost luminescent white sea ice. We looked around to see if there was a way to walk around this first lead in the ice, but it extended as far as we could see in either direction. This indicated it would be a very long detour. The easiest choice was to swim across.

On the way to the Geomagnetic North Pole to officially begin the expedition, we had travelled with a group of Inuit hunters, and one of them had told us a story along the way about a walrus that took his brother. They had been out hunting in kayaks and the walrus came up behind his brother, grabbed him and dragged him down into the water. He saw it happen, but because the walrus was behind his brother, he couldn't attempt to shoot it. He never saw his brother again. It was a tragic story. And right now, as I stood looking at the dark mirror of the Arctic sea separating us from the adjacent stretch of sea ice, all I could think about was a walrus popping up next to me as I swam.

We put our immersion suits on. These are designed with precisely that—immersion—in mind. Each suit is completely waterproof, with gloves and feet, and you step into it and zip it up around you. It goes over the top of everything you're wearing. They are also bright orange for some reason. The only parts of you poking out are your eyes and nose, which you need to keep above water, or risk filling the suit and sinking.

Once the suit is on, you carefully lower yourself into the water, which feels like being shrink-wrapped in cryogenic liquid as the daggers of cold press against you. Rob went first. He attached himself to a rope, so if anything happened, I could quickly pull him back to shore. The pulks we had were designed to float, with enclosed sides and bottoms, so Rob swam, pushing his pulk ahead of him.

In the first few days of the expedition, we had to cross leads like this four or five times. The suits kept us dry, but surrounded by icy water we only had a few minutes to swim awkwardly to the other side, manoeuvring our floating sleds along with us. Climbing out the other side required surprising agility, especially when bogged down in the soupy semi-ice, which tended to form a transition zone between the hard sea ice and the open water. It was often necessary to have ice stakes on hand to drive into the ice as claws to gain purchase and drag ourselves out.

To an extent, we had come prepared for these crossings. We knew, of course, that with sea ice there's always the chance of it cracking, as the tides and motion of the ocean create movement. But we weren't prepared for the speed at which the ice was fracturing that year. It really was unprecedented.

We had researched satellite imagery from the previous 10 years, examining the area around northern Greenland to understand the annual pattern of when the ice breaks up. Our plan was to travel to the edge of the ice to board a yacht for the next section south, so we needed to determine when and where we should aim to meet the boat. It had seemed clear we could expect the ice to be stable until June. But it was now only April, and our planned path south was rapidly disappearing. It turned out that 2007 would see the fastest sea ice melt in recorded history, and since then it's gotten even worse.

The leads in the melting ice weren't the only obstacle slowing our way. Just as the sea ice can crack apart, it can also crumple together, which presents an entirely different problem. When two large pieces of ice came together with enormous force it creates mounds of precariously stacked ice blocks. The two of us would have to climb over each mound, wrangling one sled at a time, desperately trying to avoid tipping it, toppling it or tearing it open on the jagged terrain. On each ridge we scanned the horizon for signs of the path of least resistance and the pans of flatter ice for easier travel.

It didn't take long for us to realise that our plan to skirt around northern Greenland on the sea ice simply wasn't going to work, and we reasoned it would be safer—and hopefully quicker—

to go over Greenland itself. We were tracking satellite imagery which showed the state of the sea ice ahead of us and saw that to the south of us there were breaks in the ice kilometres wide. It's one thing to swim 20 metres, but it would have been suicidal to try to swim a kilometre with the kit we had.

To continue south without traversing sea ice, we had to first cross one of the vast peninsulas that jut out from the western edge of Greenland. After we left the sea ice, we had to drag our pulks up the glacier through the moraine. It was quite steep, with more snow than ice, and our pulks got bashed around on the rocks beneath the snow. It seemed to become harder and harder to haul them along.

On the second day of this, I was walking behind Rob when I noticed a strange groove his sled was leaving in the snow. We stopped for something to eat and discovered that the rocks had been scratching away at the base so much it had basically turned his sled into a snow plough. That helped explain why we were finding it all so difficult!

At this point of the trip, making our way up the moraine and onto the glacier, we had a period of about a week when we were only covering four kilometres per day. Even though the steep sections only had a gradient of six or seven percent, with the slippery snow and ice, and the heavy sleds, it was a lot. We knew at that rate, with around 40,000 kilometres still to go, we would never make it.

It was tough not to get frustrated and deflated about this. We were enduring temperatures that would freeze a pan of boiling water in minutes, not to mention our fingers and toes. The condensation from our breath would freeze to the first thing it encountered, usually our eyelashes, causing our eyelids to fuse uncomfortably together. We had storms that confined us to our tent for days, whipping up a blizzard so thick that if you stretched your arm out of the tent door you could barely see your hand.

But once we—eventually—got onto the top of the glacier, we were able to move more quickly, skiing across the ice sheet for about a week. Then we headed down the other side. On the southern side we came across a crevasse field, a labyrinth full of huge, deep cracks in the glacier as it tumbled down the mountain. Not being able to see a clear route through, we got caught in the maze.

It was quite terrifying trying to navigate the steep slope with 100-kilogram sleds. On this particular day conditions were awful too, with powerful winds. Once these winds hit the glacier, they blow the snow away, leaving slippery bullet-hard blue glacial ice to walk across. We put in some ice screws and secured our sleds to them, so that we could go for an explore without being pulled around.

There were times for each of us that a leg would punch right through a snow bridge and hang into nothing, and we would realise we were on the snow-covered lip of a crevasse. We were roped together, which provided a level of safety, but thankfully neither of us completely fell through into the gaping caverns below. Even travelling as a roped pair is fraught with risk. If one of us were to slip into a crevasse, the other would need to work quickly, using ice screws to set up an anchor point and pulley system, to have any chance of a successful rescue. Easier said than done wearing thick gloves when removing them would mean frostbite in a matter of minutes.

In the end, we found what seemed to be a route through the crevasses. Once we were through the worst section, we decided to take advantage of being on a slope and use our sleds as toboggans. This was for safety as well as speed. When you walk on crevasses concealed by snow cover, all your weight is on the small, concentrated points beneath your feet, which makes you more likely to fall through. But we figured using the sleds to slide down would firstly allow us to spread our weight over a larger area, and secondly allow us to travel down the glacier with some momentum. Combined, these two factors would hopefully take us over any weak parts of the glacier without falling through. We lay headfirst like torpedoes on the pulks, so we could dig our crampon spikes into the snow and ice on either side to help with slowing down or steering. We took off, trying follow the hopefully safer route down the more compressed inner edge of the glacier, where crevasses would be less likely.

In some ways it was a really scary day, but in other ways it was really enjoyable to be—for want of a better word—bobsledding down these glaciers to get back to the sea ice on a frozen fjord on the other side. Looking back, we rode our luck as much as our sleds that day. We were fortunate to make it through unscathed, but as any mountaineer will tell you, those are the days that stay with you.

After that experience we had two revelations. One was that it was incredibly slow and hard going to take heavy sleds up and over the top of glaciers. The second was that we were ill-equipped to navigate the glacial areas of inland Greenland. We had only very, very basic maps and GPS. We were relying on gut instinct and almost just following our noses. We realised it would be really beneficial to have some local knowledge and experience to draw on as we attempted to navigate this unfamiliar terrain.

We decided that when we got to the most northern settlement, we would attempt to find some Inuit hunters heading south and ask if we could travel with them by dog sled. At that time of year, lots of people from the settlements along the coastline would be travelling our way. It was just a matter of finding a group willing to let us tag along.

Having pulled our sleds across the sea ice and over glaciers for approximately three weeks, we arrived in Qaanaaq, the northernmost major outpost in Greenland. We were lucky to find some Inuit with plans to go on a southerly hunting trip soon after our arrival and we quickly restocked essential supplies and organised the next stage of our journey.

Rather than a mode of travel, dogsledding is more of a lifestyle. Traditionally, hunters have headed out onto the ice for many weeks, even months at a time, living self-sufficiently with a pack of 10 to 15 dogs and what they can carry on their sleds. So even though we were back on sea ice, we were with people who were absolute experts in navigating this mode of travel.

Prior to modern times, when wood and plastic became available as construction materials in the far north of tree-less Greenland, dog sleds were constructed from whale bones and frozen cuts of meat, lashed together with cord made from seal hide. Blubber from hunted whales and seals was used for fire and cooking, and the meat fed both hunters and dogs. If the hunters were caught in bad weather, or became isolated in the harsh environment, they could survive by literally eating their sleds. While this practice has long since disappeared, travelling with Inuit hunters is nonetheless a remarkable experience in how to live and survive in this formidable landscape.

There's a common tale in the high Arctic that you could land a Boeing 747 on 10 centimetres of sea ice. It might not be technically accurate, but the premise holds that if you have sea ice 10 centimetres thick, it's extremely strong. Due to the salt content, sea ice is highly flexible, not brittle like freshwater ice. If you bounce up and down as you walk across the sea ice, you can feel the flex, similar to when you walk across a swinging bridge. And if you jump up and down rhythmically, you can feel the resonance in the ice. But you know the ice is strong, so it's not something that you worry about.

As we travelled further south, we noticed more flex in the ice, and we heard the Inuit hunters commenting that the ice was melting earlier than they expected. We also saw sections with melt on top of the ice. When the covering snow melts, it creates a darker surface, and the albedo (the measure of how light, or reflective, a surface is) is decreased, which allows more heat to be absorbed in that patch. So, you can have a situation where you have melt coming from beneath if you have currents bringing warmer water, plus melt from above with this slushy layer of melting snow and ice on the surface. On top of this, sea ice isn't a uniform thickness to begin with; the thickness depends on how it froze, if it's been shaped by currents and if it froze around fragments of glacial ice.

After we encountered the slushy surface ice a few more times, the Inuit hunters became a little concerned because it was a sign the ice was melting and could start to break up. But, based on their own previous experiences and knowing how early it was in the year, they decided maybe it was just a bit of an anomaly. Perhaps it was just one patch where that was happening? And perhaps the conditions would improve as we continued? But we were aware the ice wasn't in the best condition compared to previous years, and we all started to have conversations about how realistic it would be to get much further south.

One day, after dog-sledding southwards for about two weeks, we had an especially sunny, windless and less chilly than usual day, perhaps a mere -15 degrees Celsius. Whilst dog-sledding along, Rob removed his now-unnecessary bulky outer gloves and placed them on the sled beside him. Since setting off earlier that morning, the texture of the ice had again been a little concerning. A thin layer of slush sat on the ice's surface, bogging the sleds' runners down and impeding their progress like a gluey sludge. The sea ice itself had a greater flex than usual and slow wobbling vibrations rippled outwards if you walked on it. Crossing a slightly rougher section of ice along a joint between two plates, one of Rob's gloves fell off his sled onto the ice. A few minutes later, when Rob noticed it was gone, he stopped the sled and got off to walk back and pick it up.

As he bent down to grab it, the ice gave way beneath his feet, plunging him into the arctic ocean beneath, the piercingly cold liquid instantly penetrating his clothes. On the way in, he smacked his head on the ice, knocking himself unconscious and at the mercy of the life-sucking water. Looking back to see why Rob had stopped, I witnessed this happening from a couple of hundred metres away and knew I needed to get to him as swiftly as possible.

Hoping that the sea ice wouldn't swallow me too, I ran back towards him with the lightest steps I could muster. My heavy clothing weighed me down, and it must have taken three or four minutes to reach him. I called his name, but he didn't respond. Lying down and stretching out to spread my weight across the thin ice, I grabbed his hood and then his armpits and slowly began to heave him out of the water, his now-drenched clothes doing their best to resist my pull.

When I finally had him out and lying prone on the ice, I saw that he was beginning to turn blue from the cold. I knew we didn't have much time. I started stripping his wet clothes off to stop them from drawing out his body heat and soon he was almost naked. I grabbed his sleeping bag from his sled and unzipped it beside him before rolling him into it. I removed my jacket and climbed in with him. The two Inuit hunters we were travelling with quickly pitched their makeshift tent over the back of Rob's dogsled and lit the stove inside. Once it was ready, we carried Rob into the tent. We took turns to hug him to pass on our body warmth, and we massaged the blood into his hands and feet.

'Rob?' I asked, but there was no reply.

We had trained for a scenario like this, learning the correct steps to take when someone has severe hyperthermia, but you never truly expect to be in the situation. There is something eerie and disturbing about a cold and unresponsive body that makes the situation harder to take in, and you wonder about the futility of doing anything. We continued to get as much warmth into Rob's sleeping bag as we could. We heated water on the stove, creating makeshift hot water bottles from our drinking bottles to tuck into the sleeping bag. It was still not clear if they were having any effect, though. Outside, the dogs ruffled and barked as if they were aware of the situation's urgency. The hunters watched Rob intently for any signs of movement.

Then, in a barely audible croak, Rob began to mumble. Colour started to return to his face. We made some lukewarm sugary water and began to feed it to him in tiny sips. As the sugars trickled into his stomach, little by little, his eyes became wider and regained their sparkle and he began whispering brief responses to our questions.

We called for a rescue helicopter to evacuate Rob and cleared an area for it to land. There aren't usually many helicopters hanging around in northern Greenland, but a handful of villages in the north are serviced by them, so the emergency services were able to divert one that had been doing a commercial flight to a town a couple of hundred kilometres away. We loaded Rob in and it took us to a medical facility in a little village. When I had pulled Rob from the icy sea, I wasn't sure whether he would survive or not, but over the next few days he made a full recovery.

At that point a lot of people back in the UK who were supporting us started saying, essentially, 'Good try, guys, but I guess it's time to come home.'

We had only been out for six weeks, and they thought that was it. However, Rob emerged even more determined to finish the expedition. Even though it's not possible to link a single incident to climate change, in our heads it felt quite obvious that the early sea ice melt that had made our journey so difficult, and nearly ended the expedition, was caused by climate change. Had the ice not been thinning and breaking up as early as it was—and so far north—this wouldn't have happened. Paradoxically, the accident just filled us with determination. It was more important than ever to continue the trip and continue telling that story.

After just over a week in the village, our expedition yacht arrived, and we continued on, sailing south through the Davis Strait and onwards to New York. Ultimately, the trip was a success and 11 months later we sailed under the Sydney Harbour Bridge, having had close calls with icebergs, cycled through rainforests and deserts and been flipped upside-down by 30-metre-high waves in the Southern Ocean. We had travelled from the Geomagnetic North Pole to the Magnetic South Pole.

Once we got back from the expedition, we were invited to several schools that had followed the journey. We also received a lot of really nice letters from students saying how interesting they had found the expedition and that it had brought their classes to life. Seeing that it inspired others was very rewarding. But in many ways, the trip ultimately inspired me with climate and environmental issues more than I could have ever inspired anyone else. In addition to the melting sea ice, we witnessed the incredible pace of melt of Greenland's glaciers, cycled through exceptional heatwaves and drought in the southern US, saw the clearance of rainforest in Central America, the impacts of overgrazing in southern Argentina and had to be on watch for unseasonal icebergs in the Southern Ocean following the break-up of the Antarctic Ice Shelf.

It was these experiences that, despite my lacklustre performance in the last years of high school, inspired me to study climate change formally. I first enrolled in a geography degree, then undertook a PhD in Earth and Environmental Science, investigating the feedbacks between agriculture, wind erosion and biogeochemical cycles. I investigated the effect of contemporary and historical agricultural practices on dust production and marine fertilisation—which is hypothesised to have a significant impact on the climate through sequestering carbon from the atmosphere to the ocean floor.

During my PhD, the skills from the expedition came into play when I needed to travel to polar environments for my research. I undertook a trip to South Georgia, a glaciated sub-Antarctic island, best known as being the place where Ernest Shackleton arrived after rowing from Elephant Island following the loss of his ship, the *Endurance*, in the pack ice.

We sailed to South Georgia—you can't get there any other way—and when we arrived, we pulled our sleds up the glacier to reach a plateau. Once on top of the glacier, we were able to drill down into a stable area to get an ice-core. While it was different to Greenland, the skills I had learnt there meant I felt much more comfortable in the polar environment and was confident I could conduct the research safely. In reality, there aren't an especially huge number of scientists who have both the interest and the polar survival skills to undertake this kind of research.

I was there to find out how much dust was caught in the ice. My PhD was investigating whether there has been an increase in dust as a result of human land use expansion and the intensification of human agriculture. This is thought to have happened at different times in different parts of the world, but I was looking at South America, focusing on the period when Europeans arrived and started cattle ranching, particularly with sheep. Dust emissions from South America are particularly consequential because downwind of South America is the South Atlantic and the Southern Ocean, and this area of the ocean is iron-limited—it's what's called a high-nutrient, low chlorophyll zone. It has lots and lots of nutrients and is really good for supporting life, but it is deficient in iron—and iron is a vital ingredient to allow life to grow, especially for phytoplankton.

There's a hypothesis, the 'iron hypothesis', that suggests that when we look at increases and decreases in carbon dioxide over the past thousands of years between glacial cycles, one of the possible mechanisms that could be helping drive the amount of carbon dioxide, and the earth going in and out of glacial periods, is the amount of dust. How does that work? Dust essentially carries soil to the ocean and provides iron to these areas of ocean, fertilising them and allowing phytoplankton to grow. As the plankton grows, the hypothesis is that they absorb carbon dioxide from the atmosphere into their bodies. When they die, this is deposited on the bottom of the ocean, effectively creating an immense natural carbon sink.

This idea of dust fertilising the phytoplankton, and the relationship between dust and climate, are still relatively poorly understood. In the time since the industrial revolution, when we've been producing more carbon dioxide, we've also been producing more dust which may, unbeknownst to us, have been resulting in more carbon dioxide being removed from the atmosphere at the same time. If it turns out we've been removing carbon dioxide from the atmosphere in a way that contemporary climate models haven't properly captured, that could have important implications. It could mean that if we are able to improve land use practices and reduce dust emissions (and reducing soil loss is very important), then we may unsuspectingly reduce a carbon capture mechanism that we have been stimulating. That may result in a more significant increase in carbon dioxide in the atmosphere than models predict.

I wanted to discover whether there was a contemporary relationship, or correlation, between dust events and an increase in phytoplankton growth in the Southern Ocean. Certain species of phytoplankton release a unique sulphur compound when they grow—they are the only possible source of this sulphur compound in these areas. So, my goal in taking the ice core was to effectively chop it into little slices and to look at each slice to see if, when there is more iron-rich dust, is there also more of this biogenic sulphur compound? This turned out to indeed be the case, which suggests there is quite a good relationship between increased dust and phytoplankton growth in that area.

However, we also found that unprecedentedly warm temperatures just 18 months before we drilled the ice core had caused meltwater to percolate down through the ice, literally washing away years of critical records that would have allowed us to better measure this relationship. Just like in Greenland, we had arrived a year too late. The changing climate that had caused unprecedented sea-ice melt in the Arctic in 2007 was now contributing to off-the-scale temperatures in the sub-Antarctic in 2015.

There have actually been some experiments where people have essentially taken boatloads of iron-rich dirt out into the middle of the ocean and dumped it over the side and studied what happened afterwards. The majority of those studies did find that there was a substantial increase in phytoplankton growth immediately afterwards. These kinds of experiments have since been banned under international treaties. They were seen as geo-engineering and frowned upon.

But even if the relationship between dust and phytoplankton is established, it's not 100 percent clear what happens to the phytoplankton once it has grown. There's debate in the scientific community around to what degree phytoplankton dies and sinks to the bottom of the ocean (and stores the carbon) and to what degree it decomposes or is eaten by other sea life that eventually dies and decomposes (and releases the carbon back into the atmosphere). This means we can't measure how much carbon dioxide phytoplankton removes from the atmosphere with any certainty.

Despite these unknowns, this research highlighted the complexities in the Earth's system and how the flow-on effects of one change has the potential to influence something seemingly unrelated in a major way. The climate system is full of these feedbacks, some positive, some negative, and understanding them is key to being able to predict how the world may change as a result of increased greenhouse gas concentrations and global heating. This is only one of the many areas of research to better understand the Earth's systems and the impact that human activity has had on their operation, particularly during the industrial period.

My life has changed since my experiences on the ice. I've been having a few less year-long adventures recently. Instead, I have been working with businesses and industry to help them understand and reduce their carbon emissions and prepare for the risks posed by atmospheric carbon dioxide levels that haven't been seen for millions of years. I'd like to think that my life will come full circle and one day I will visit the far north again; that I will get to meet the Inuit hunters who helped us, and the other friends we made.

On more pessimistic days, I fear that the stunning ice-clad landscapes of Greenland will have changed beyond all recognition and that the skilled hunters will have had to swap their dog sleds for motorboats. If we do not act as a society to reduce our greenhouse gas emissions to zero, this is the stark future they face. However, I hope that the momentum I see building will lead the way to a more sustainable future, where we realise that we are part of this planet and do not somehow operate outside of the natural systems that have produced us and kept us alive so far.

OUT OF SIGHT... BUT NOT OUT OF MIND

BY
OLIVIA JOHNSON

Olivia Johnson graduated from the University of Tasmania with first-class honours in Marine & Antarctic Science in 2018. Between undergraduate and honours, Olivia undertook an internship in the Maldives as a marine biologist and post-honours was fortunate enough to be named the 2018 Our World Underwater Scholarship Society's Australasian Rolex Scholar, where she spent a year immersed in all things diving around the world. After this Olivia worked full-time as a research assistant for the Institute for Marine & Antarctic Studies, with her role primarily focused on SCUBA-based reef surveys. Recently returning from a voyage south as a Krill Biologist, Olivia has now begun her PhD, 'Safeguarding threatened reef species', at IMAS, UTAS. Her major interests lie in marine ecology and conservation.

I was fortunate enough to be raised by an ocean-loving family by the sea in Hobart, Tasmania. When my brother and I weren't at the beach with Mum fossicking through rock-pools or sifting our way through a clump of seaweed to see what creatures we could find hidden amongst it, we were helping Dad, a commercial diver. We would hose his boat down and see what would wash out of the scuppers after a day on the water harvesting the local sea urchin. Family trips to Bruny Island or White Beach always saw Dad bringing back a feed of abalone, crayfish and scallops that both my brother and I would investigate until Mum asked for us to bring them inside so she could cook them. Summer days were spent with wetsuits and snorkels on, swimming along the reefs to see what fish we could spot.

With a childhood like this, it was no surprise that when I was given the chance in high school to learn to dive, I jumped at the opportunity. Within a year, I'd completed my Open Water, Advanced Open Water, and a Rescue Diver course, and I had my heart set on completing a marine science degree. I'd applied to the University of Tasmania and received an offer in a Bachelor of Marine & Antarctic Sciences. For me this was even more exciting because this was the first year that the degrees had been combined, and while my first choice was marine science, my second choice was Antarctic science. Hobart is a gateway to the Antarctic and seeing the big red ship (the *Aurora Australis*) in port or heading out to sea was common. During my school years, Antarctica had always come into the picture somehow. I was thrilled to be starting a degree in both!

The first year was a tough one—the course work was harder than I'd expected and it wasn't as marine-based as I had hoped-but I made some fantastic friends. We all passed and were excited for the second year where we would start to get into the more specialised marine and Antarctic sciences. I also enrolled in a Scientific Diving summer school course with the university at the end of my second year. With my diving qualifications and a few years' experience diving recreationally with my Dad and my friends, I knew I wanted to apply these practical skills to the knowledge I was gaining at university.

Mid-way through my second year, on a cold winter morning in July, Dad woke me very early to head up the east coast and go scallop diving. I loved a sleep-in, but I was very excited to be going scallop diving for the first time with Dad, especially after watching him bring them home for years as a child. Most of our recreational diving prior to this had been along rocky reefs, me taking in the wonders that lie beneath the surface whilst Dad busily buried his head under a rock looking for crays. But it was the last day of the recreational scallop season and we both had our licences so we could collect 50 scallops each.

It was a beautiful morning; the water was totally flat and looked like glass. Once on the boat, we made our way to the dive site between Maria Island and the mainland to find a nice patch of bed that Dad had previously dived on. We anchored the boat and, once we were geared up and happy, Dad started the compressor. We jumped in and headed for the bottom, using the anchor line as our reference. I remember the water being a crystal-clear blue. I couldn't believe we could almost see the bottom after only descending a few metres.

Once there, Dad started to show me the best way to find good-sized scallops buried in the sand. I was so excited to be diving in a new type of habitat and seeing all the new marine creatures along this sandy bottom. We both filled a bag each and headed back up to the boat to measure the legally sized ones and counted out our 50 each. We were short by 14 proper sized scallops. I was freezing. The water was only around 11 degrees and my wetsuit had a hole in it. So Dad moved the boat a little bit away from where we had been, just to see if there was a better patch nearby, and decided to jump back in to grab the last few scallops.

Dad re-anchored, geared up and started the compressor back up, which roared loudly. Just as he was about to get back into the water, I yelled, 'See you in five!' As he rolled backwards into the water, he gave me a wave.

You know when you get a gut feeling that something isn't right? Around 20 minutes after Dad had jumped back in, that feeling began to strike. I'd spent my time sitting in the boat busily trying to shuck an oyster I'd picked up along with my scallops earlier. I was using a fish filleting knife and understandably it hadn't been going well. I had been intermittently glancing up to look at where Dad's hookah line (a breathing apparatus that linked him to the boat) was stretched out. Sometimes sitting on the boat versus being underwater, your perception of time is a bit warped. But after checking my phone to see the time, something just didn't seem right. I decided to ring Mum as she had been out with Dad the weekend before and sat on the boat while he dived. Thoughts had crossed my mind, but I didn't want to think the worst before I knew. Mum said it didn't sound like Dad to be down for that long and that I would need to get back in the water to look for him.

I did as Mum said and I found Dad. He had been attacked by a great white shark.

I went into full fight or flight mode. I think in my rational head I knew that Dad was gone, but obviously I was going to go back to the boat to get help as fast as I could. Because his hookah line was stretched all the way out, I took hold of it as I swam back to the boat just as some kind of reference in the water. But as I swam I suddenly realised that with my heart racing, holding on to the hookah line would create a vibration through the water that might attract the shark to me instead. I instantly let go and just swam back to the boat as fast as I could.

The rest of the day and weeks afterwards were a whirlwind. But in the aftermath of all the events that happened that day, I vividly remember thinking to myself, 'I can't go back to university, I can't be a marine scientist now.'

I knew going back to uni people were likely to tip-toe around me and I didn't want to be that person who was looked at differently. I also didn't want this experience to define me. I had been really loving university and wanted to learn as much as I could, but I knew it was inevitable that sharks were going to come up and I didn't know how I would be able to cope with that.

While I did return to university after a few weeks, it wasn't until December that year that I felt able to get back into the water. My boyfriend at the time was also a diver, so we decided to just go for a dive at a local beach to slowly start to get used to the water again. Neither of us had been in the water since my dad's death and we were both very on edge. I remember constantly checking over my shoulder and having the absolute life scared out of me when a small draughtboard shark swam past. But we were in a very shallow marine reserve and it was a lovely calm day, with clear visibility and lots of small critters around the kelp forest, which also helped me remember why I loved diving so much. Despite my nervousness, I realised that this was something I could still do.

In fact, science and research helped me process what happened to my dad. Just thinking about it, it really didn't seem real. But researching information about great white sharks and other shark species helped me to process what had happened to Dad and helped me to make the decision to finish my degree. A hunting shark mis-identifying a person for a seal made sense. Dad gone did not make sense. The more I researched, the more the guilt I felt for not being able to help Dad that day started to slowly chip away.

Six months after Dad passed, I enrolled again to do my Scientific Diving course. I began the course after a little family holiday with my mum and brother in Queensland. I did some beautiful coral diving on the Great Barrier Reef and was reminded again of how much I loved being in the ocean, and how connected to my

Dad I felt being in it. The glorious warm waters were bustling with life and the coral reefs I saw were thriving! It wasn't my first time diving tropical reefs, but it cemented how much I wanted to make a career out of it.

The Scientific Diving course was my third time back in the water since the tragic scallop diving trip. As much as I had to concentrate on all the new tasks and assignments I was given, I still couldn't help but look over my shoulder every few minutes.

Through my experiences learning about sharks and getting back into the water, something else crystallised for me. I had learned from the public exposure I'd had from Dad's accident, how easily the media can take, manipulate and sensationalise the truth to create a story that not only sells, but really influences people's views. While the sensational reporting on my father's death was upsetting to my family and I, there was more to it than that. Sharks themselves are frequently misrepresented. In my pursuit to find out as much as I could about sharks, some of the articles and information I came across online left me horrified; it was the kind of material that would instil fear and encourage pure hatred for sharks. When people absorb attitudes like this, it's not their fault per se—people rely on platforms such as the media to shape their knowledge and understanding of the world. Not everyone has the opportunity to get in the water and experience sharks, or any marine life at all, first-hand.

This extends beyond marine life too. If the general public can't observe something for themselves, modern media has a unique opportunity to accurately educate and inform. This includes really important and topical issues such as climate change, which often are 'out of sight, out of mind' to those who do not already know or understand. And one of the biggest topics that came up throughout my degree, across all subjects, was the effect climate change was having on our oceans—changes that are invisible to most people in their day-to-day lives.

This became particularly apparent in 2016, my third year of uni. I began working with a dive team investigating the effect of the long-spined sea urchin along Tasmania's east coast. This urchin, *Centrostephanus rodgersii*, has shifted its native range due to a combination of factors, but one of the major ones being the extension of the East Australian Current down the east coast of Australia to Tasmania, bringing warmer waters and the larvae of the urchins. The warming Tasmanian waters had already seen a significant loss of the giant kelp forests around the state. The range-shift of new grazing species like the long-spined sea urchin over the last 40 years has now had detrimental impacts to other macroalgae—also known as kelp species—along Tasmania's east coast.

Some of the local Tassie reefs that we came across during diving, or from undertaking towed video surveys, had been completely wiped out by the long-spined sea urchin. But when I spoke to locals, they had no idea what was happening in their own backyard. I think that hit the nail on the head in terms of 'out of sight, out of mind' for me. It showed how easy it is for people to be disconnected from what is happening so close by.

Soon though I was headed further north to study the effects of climate change in tropical waters. After finishing my undergraduate degree, I was very lucky to land a six-month job opportunity as a marine biology intern in the Maldives. Coming from Tassie, the journey took a couple of days, but once I arrived at my new home, One and Only Reethi Rah, in North Malé Atoll, and met my new boss and fellow Australian, Kylie Merritt,

I could not wait to jump in the water and look at where I would be working.

The first place Kylie took me to was Turtle Reef to meet some of the resident Hawksbill Turtles and show me the reef that we would be taking guests on daily tours. One moment that has stuck with me to this day was my confusion as I jumped in the water and looked around at the reef. I remember there being lots of fish that caught my eye, but something didn't look right. There weren't the bright colours that I remembered from the Great Barrier. I knew I was in a different country and a completely different part of the ocean, plus I didn't want to ask a stupid question on my first day on the job, but I just couldn't put my finger on what it was.

So, when I stuck my head up out of the water, the first thing I asked Kylie was, 'What's wrong with the reef?'

Kylie looked at me and said, 'The reef is dead Liv.'

I was shocked. I knew the reefs in this area of the Maldives had suffered catastrophic bleaching in May of 2016, but I simply hadn't expected the bleaching to have been so severe. Nearly six months on, the dead coral was covered in a green filamentous algal growth and the reef was mainly dominated by a green zoanthid, the only thing that really gave the reef any colour. This was the first time that I had truly seen the effects of climate change on tropical reefs.

The next year, I was awarded the highly competitive Rolex Scholarship, which allowed me to travel extensively and gain experience across many different marine environments. In a 12-month period, I conducted 240 dives, spending 200 hours compressed, which is the equivalent of eight consecutive days underwater, went on 71 flights, spent 84 days on ships, visited four continents and 15 countries and experienced temperatures ranging from -10°C to 40°C.

During the scholarship year, I quickly found myself at the opposite extreme to the Maldives, in the Antarctic waters, seeing how climate change affected cold water ecosystems in different ways.

In January 2018, I set off on a seven-week voyage on the CSIRO Marine National Facility Vessel, RV *Investigator,* as one of 28 scientists on the most comprehensive scientific study ever undertaken on the world's largest animals, Antarctic blue whales and their food source, Antarctic krill.

The voyage, led by the Australian Antarctic Program, was called the ENRICH Voyage, standing for 'Euphausiids and Nutrient Recycling In Cetacean Hotspots' (euphausiids are krill; cetaceans are whales). This voyage was made up of different teams including the krill team, the whale observers team, the passive acoustics team, the active acoustics team, the trace metal team, the unmanned aerial systems team, the biogeochemistry team, plus the media team onboard to film for an upcoming documentary series.

It took us approximately seven days to reach our destination and the seas were rough to say the least. The swell was 10 to 12 metres, with the prevailing winds hitting us in each of the latitudinal regions (the Roaring Forties and the Furious Fifties), leaving most of us in our bunks for at least a few days until we got our sea-legs.

The Voyage was led by Chief Scientist Dr Mike Double and Deputy Chief Scientist Dr Elanor Bell. I would make up one of seven members of the Antarctic krill research team, which was undertaking a range of research aims including capturing krill swarms in 3D, utilising state-of-the-art echo-sounder technology for the first time on an Australian research vessel in the Southern Ocean. Before this voyage, very little was understood about the various types of krill swarms. All of this information helps to make better policy decisions for an incredibly valuable food source for not only whales, but many of the charismatic (and non-charismatic) animals in the Southern Ocean. And where we found krill, we usually found whales; Antarctic blue whales almost exclusively eat krill. In one day, an adult blue whale can eat up to 4 million krill, more than three tonnes! In order for them to do this, they must feed in areas where they can find high concentrations of krill.

The krill team was split into a day and night team (I was on the night team), working around the clock to understand the patterns and behaviours of Antarctic krill, *Euphausia superba*. On our way south, and on the way home north, we towed a CPR—a Continuous Plankton Recorder. This piece of scientific equipment has remained largely unchanged since its original design in the early 1900s and is central to one of the longest running marine biological monitoring programs anywhere around the world. The CPR works by filtering plankton from the water over long distances (in our case 450 nautical miles) on continuously moving bands of filter silk. This then gives us a sample representing each of the plankton groups (both zooplankton and phytoplankton) over a continuous period of time—ingenious! This process is conducted on every voyage south by an Australian vessel.

Once we hit south of 60 degrees, we were on the hunt for krill swarms. We would be deploying a net, known as the RMT (Rectangular Mid-water Trawls), where we would take a sub-sample of a swarm to get a better understanding of the biology of the animals that made up the swarms. Dr Joshua Lawrence, the acoustician onboard, used the multibeam echo sounder to help direct us to the swarm to take a sample, as well as to 'paint a picture' of the size, shape and density of krill swarms that were being observed in three-dimensions.

We were only aiming to take small samples of the krill, which we were using for multiple purposes. This included biological samples that were snap frozen for genetic and further analysis back in Hobart and collecting gravid female krill (female krill full of eggs) in order to capture the fertilised eggs. We would be rearing the eggs through their multiple larval stages, bringing them back to Hobart to the Australian Antarctic Division where experiments would soon be undertaken to get a better understanding of the influence of increasing CO_2 levels on the larvae. This is an extremely important question because of the fundamental role krill play in the ecosystem and food-chain.

Research at the Australian Antarctic Division shows that if we continue on our 'business as usual' approach to carbon emissions, which leads to ocean acidification, by 2100 there is unlikely to be suitable habitat for krill reproduction; only 50 percent of krill embryos will hatch across most of the Southern Ocean and by 2300 the whole krill population in the Southern Ocean may collapse. This would have catastrophic consequences for the entire Southern Ocean ecosystem, as well as the global ecosystems it links to.

Krill is what is termed a 'keystone' species—in one way or another this species links the entire food web together, from the krill eating the small, microscopic plants termed phytoplankton, to the krill feeding the ocean giants like the whales and seals. A complete population collapse will result in both bottom-up and top-down trophic cascades—the ramifications would have major implications for the global biodiversity that humans so deeply depend on. This presents just another example of the significant consequences climate change is having on our world.

When we were not busy in the lab, we also got out on deck as much as possible so we could take in the incredible views of the icebergs and see the array of wildlife the Southern Ocean and Antarctic has to offer. It didn't matter what time of the day it was (or whether we were supposed to be asleep or not), when the ship encountered its first group of Antarctic blue whales, every single person was out on deck. Three enormous blue whales came right next to the ship. They even swam under the bow.

I remember this so vividly, my first time seeing the world's largest animal so gracefully breaking the water's surface and blowing the most enormous stream of water. Then, just as quickly, they dove straight back down under the ship. The encounter lasted for many minutes, with the whales popping up around the ship in multiple places. It was an incredible moment and a beautiful memory to have. For years I had learned about the theories behind climate change and the effect it is having on our oceans. I knew that the oceans were warming, and places like the freezing polar oceans were providing a carbon 'sink' for all of the excess carbon dioxide humans are producing, with large ramifications for ocean chemistry and all ocean life. But that moment, watching those incredible animals in such a wild and remote place, made the desire to study and gain as much knowledge as we possibly could even more urgent.

For me, the next step had to be diving with white sharks. Partly it was about my journey to finally feeling completely comfortable in the water again. The decision also came from the knowledge that for most people the vast majority of the underwater world is hidden and it is very hard for people to care about something they don't know about or understand. The stigma that surrounds predatory ocean species such as white sharks just instils further fear in most humans, making them even less inclined to want to learn about the various life forms in the ocean, not just one species.

So, over the New Year of 2018–19, I set off on the most personal and testing journey of my scholarship year. Deciding to dive with great white sharks was one of the most difficult experiences I have ever chosen to do, but quite possibly the most rewarding. I went with Rodney Fox Expeditions in South Australia, a research and eco-tour company that spends a lot of time around the Neptune Islands, off Port Lincoln, a breeding and habitat ground for the species.

I knew most likely this would be a really great learning and healing process. I knew from the research that I had done on sharks, particularly white sharks, that mistaken identity and wrong-place-wrong-time for some people, such as my dad, was often the cause of accidents with humans. Of course, there are other reasons, but this tends to be the main cause. This top apex predator is a protected species in Australia. Being given the opportunity to potentially be in the water and diving with one would be a privilege, particularly as they have been around much longer than humans. It would also be a great opportunity to feel safe diving in the cage, while observing a white shark in its natural habitat swimming amongst the reefs, seagrass beds and on the sand-edge. And to try to create new, positive memories of these animals.

The effects of climate change on our oceans is impacting many species. More and more species are shifting their geographic ranges due to temperature changes. This is something that is being observed in species globally, particularly marine species that are shifting their range poleward as the water warms. Among them are many sharks that scientists are observing moving into waters they weren't previously commonly found in, sometimes hunting prey species outside their normal diet, creating disruption in the ecosystem food chain.

I was extremely fortunate to be joined by Rodney Fox's son, Andrew Fox, who owns and runs the company. We left port and headed out to the beautiful Neptune Islands, about a four-hour journey from Port Lincoln. Luckily for the guests on the first expedition I participated in, as soon as we arrived at the islands we were greeted by sharks. Andrew and I were in the last group to enter the water.

It was twilight on the surface, and nearly dark underwater. I'm not going to lie: the first dive was very eerie. As we made our way to the seafloor in the bottom cage, you could feel the presence of not just one, but multiple sharks. We were surrounded by five enormous white sharks and it was hard to know where to look. They truly are an incredible animal. I had my camera, but didn't know where to point it, so I used this dive to take in what I was seeing and what I was really doing.

I knew we were safe inside the cage, but the eeriness of nighttime diving and the fact that it was so dark as we descended to the ocean floor made my heart race more than I was anticipating it would. Once we spotted our first shark, the sheer size of the animal made me again appreciate why these animals are in a league of their own when it comes to ocean predators.

For most of our dives on the trip the cage would hang just above the thick, lush seagrass beds of the seafloor. There were often large schools of fish swimming through the seagrass and just adjacent was usually a rocky kelp reef, the kelp gently moving back and forth. Eagle rays looked like they were flying over the reefs before disappearing into the seagrass.

Over the two weeks I spent onboard the *Princess II*, I learned how an ecotour company such as this operates around protected sites and animals, as well as the different reasons people have for choosing to come and experience this. Overall, there is a general fear when it comes to sharks and getting into the water with them, so it was great to listen to other people's stories and about what had brought them to the Neptune Islands. There were many different reasons why people were onboard, but nearly everyone I spoke to told me how experiencing these great predators first-hand had changed their perception of white sharks in a positive way.

This experience helped put all my 'grieving research' into perspective. It helped give me some closure on how I had lost my dad and shed new light for me on white sharks and the importance of these animals to the ecosystem. It helped that I'd seen how the media shaped public opinion on sharks. Sharks can be seen as terrifying or as a vital part of the ecosystem. And all ocean ecosystems, whether they are polar, temperate, or tropical, interlink and are all facing adverse effects due to the influence of climate change and human activity. This will influence all humans, directly or indirectly.

Although the path to get to where I am today was not necessarily a straightforward one, it has shaped the person I am and my outlook on life. It means I look at my research through the lens of how I can educate people and communicate what is happening in the oceans, because people will only act to manage and mitigate our mark on the natural world when they understand what is truly at stake.

When I lost my dad, I didn't know if I could ever get back into the water. Pushing myself to dive again and continue my studies was probably inspired by his legacy—his determination and hard work ethic rubbed off on the whole family. He never really stopped. He would come off 12-hour night shifts and immediately head out to dive for the day. He was always happiest on or in the water. Just like I am. Being in the water brings me closer to him and so does working to help protect the oceans he loved.

CYCLING THROUGH A CLIMATE APOCALYPSE

BY

JOHN F BRUNO

Professor John Bruno *is a marine ecologist at the Department of Biology at The University of North Carolina at Chapel Hill, USA. John grew up by the ocean in south Florida and came to UNC in 2001 after a post doc at Cornell University and a PhD at Brown University where he worked on the ecology of coastal plant communities. His research is focused on marine biodiversity, particularly the impacts of climate change on marine ecosystems. Beyond documenting doom-and-gloom, the Bruno lab works on understanding how human activities alter the structure and functioning of marine food webs and what local conservation strategies are effective in mitigating these impacts. John works closely with NGOs and congressional staff on legislative issues related to ocean conservation. John is also a dedicated science communicator: he blogs at TheSeaMonster.net, writes for major news outlets including the Washington Post and New York Times, is an amateur filmmaker, and gives public talks about ocean critters and marine conservation. At UNC, John teaches Marine Ecology and Seafood Forensics and co-leads the college's QEP-CURE Program (Course-based Undergraduate Research Experience) which is developing, implementing, and evaluating the effectiveness of innovative and interdisciplinary research courses.*

E ven though it's midday, it's eerily dark outside. Since the light is so diffuse, there are no shadows on the ground. I'm biking up a steep mountain trail in central Montana and, through the smoke from a nearby fire, the sun is deep orange. It burns my throat and nostrils, but the larger effect is psychological—the smoke and haze are strangely oppressive, making me feel almost claustrophobic. I'm in Big Sky Country and normally would be able to clearly see neighbouring mountain ranges over a hundred miles away, but these conditions shrink the world down to a landscape of a less than a kilometre. People who have experienced similar climate-driven conditions in recent years often use the word 'apocalyptic' and that's exactly what it feels like—a hot, dry, smokey, orange-tinted future scene from Blade Runner. That was the summer of 2021 in the western United States.

How did I get here?

Big mountains and the high sage bush deserts of the American west aren't my native habitat. I'm a marine ecologist and most of my fieldwork is on Caribbean coral reefs or the productive, rocky reefs of the Galapagos Islands. I work underwater in tropical seas surrounded by fish, corals, and seaweeds. But I also love the mountains and alpine ecosystems and I've always been crazy about bikes. I've raced BMX, mountain, and road bikes since I was a kid. I worked as a bike mechanic and a bike messenger in my 20s. And last summer, unable to get to my field sites due to COVID, I got back into bike-packing—essentially backpacking on a mountain bike.

All my gear (clothing, food, my stove, and tent) is stored in waterproof bags and strapped to my bike's frame and handlebars. It makes for a heavy load, but also total freedom to explore and experience the terrestrial world. Peddling for hours (some days 10+) might sound gruelling, but once you adapt to it, the repetitive motion is very Zen, healing and surprisingly addictive. If the conditions are right, I can travel 50 or even 100+ miles in a day. Most days, I'll cross a mountain pass or two and a dozen clear streams. I will ride across numerous ecotones and will see an abundance of wildlife. I'm always on the lookout for grizzly bears and mountain lions. On journeys like this, your inner conversation gets mostly constrained to the essentials: the trail, wildlife, water, food (lots of pop tarts!) and where you'll camp that night. At the nucleus of our existence, that's really about all we're meant to think about.

I was riding on the 2,700 mile Great Divide Mountain Bike Route that runs from Banff National Park in Canada all the way down to southern Arizona and the Mexican border. There are some rural, paved roads, but mostly you're alone and on trails and gravel roads through the forest and across massive, open valleys. It's an incredible way to experience the landscape and to meet the people that live in it. Unfortunately, it is also how you experience climate change firsthand.

Rocky Mountain climate catastrophe

When I started the trip just outside of Glacier National Park it was 35 degrees Celsius (95°F). The previous week it was 41 degrees Celsius (110°F). In northern Montana! Summertime average temperatures in this region have increased by 2.5 degrees Celsius (4.5°F) since 1970. The science of climate attribution has advanced rapidly, and it is now known with substantial certainty that these heatwaves and the hot, dry, windy weather that promotes wildfires have become far more common due to climate change. Of course, it's not only the western US and Canada; these conditions are developing in many other regions, including Siberia, Paraguay and Australia. The new IPCC climate report makes it clear that extreme heat and fire will continue to increase as the concentration of carbon dioxide, methane and other greenhouse gases in the atmosphere grows. These gases are released by the burning of fossil fuels, deforestation and from animal agriculture. They act as a blanket, trapping heat in the atmosphere that's reflected off the earth's surface. The higher the concentration of greenhouse gases, the thicker the blanket and the warmer the planet becomes.

The smoky conditions I faced weren't unexpected. A lot of people riding the trail last summer lamented them on social media, and I've read and co-authored enough climate change reports to know that this really is our new normal. I'd thought (or hoped) I could avoid the worst by starting relatively early (on July 4), but fire season in the western US is beginning earlier and lasting longer. I felt awful for the millions of people who live with the smoke for months every summer. It's terrible for their physical and mental health. In fact, a joint statement published simultaneously in hundreds of medical journals in 2021 stated, 'The greatest threat to global public health is the continued failure of world leaders to keep the global temperature rise below 1.5°C and to restore nature.' Global warming and other aspects of climate change are impacting human health in countless ways.

Now I'm back in Chapel Hill (North Carolina, USA) and it's hurricane season. Across much of the US, it's increasingly too much water that's the problem. Category 4 Hurricane Ida just flooded Louisiana. Three days later, flooding from the storm's remnants resulted in 45 casualties in New York and New Jersey. This summer, record rainfall events have killed dozens of people in Tennessee, Kentucky, and North Carolina and it was reported that nearly one in three Americans experienced a weather disaster. Scientists have been warning about weather extremes for years as an important aspect of climate change. These catastrophic weather events will continue to intensify as greenhouse gas concentrations increase.

It wasn't always smokey on the trip. If the wind was blowing in the right direction, it would clear out the smoke from local fires without bringing in smoke from distant fires in California and Oregon. When that happened, I felt immediate relief from the psychological oppression of the smoke and haze. When I rode into Grand Teton National Park, just after rain had cleared the air, the sight of the towering, sharp and bright Tetons (which had been shrouded in smoke for weeks) gave me a truly euphoric feeling. I stopped and ate some pop tarts to extend and further enhance the moment.

But there was another omnipresent reminder of climate change: mass tree die offs caused by drought and heat. Most days I'd pass through at least one relict forest, where nearly all the trees covering a whole mountainside were dead. Some of the standing dead had also burned, creating a stark, Mordor-esque landscape. At times, my thoughts and feelings were almost clinically objective as I tried to work out what had happened—the causes and sequence of ecological events—and how to document and share it with the world.

The conifers that blanket the mountains of the west, particularly spruce and pine, are being killed off by bark beetle outbreaks exacerbated by global warming. Mild winters enable the beetles to survive through cold periods that would typically arrest their spread. And the drought and heat seem to weaken the trees, making them more susceptible to infestation. Additionally, a recent 38-year study in Colorado found that changing climatic conditions alone were the primary cause of tree mortality. These die offs also fuel wildfires and release massive amounts of carbon stored in the trees when they burn; a perverse positive feedback loop that intensifies the cycle of ecosystem degradation.

Ocean heating and coral reefs

Although it was jarring to experience climate change in the terrestrial world, I've been observing (and documenting) similar impacts in the ocean for nearly a quarter century. The mass-die offs of trees in these high elevation forests were a constant reminder of the loss of foundation species in the ocean that I study. A lot of the work in my lab is focused on the corals that build up Caribbean reefs. These colourful and biodiverse Edens of the world's tropical seas are obviously different from mountainous conifer forests, yet I see parallels in how these habitats are being degraded by climate change.

Corals are invertebrate animals in the phylum Cnidaria (closely related to jellyfishes and sea anemones). These tiny animals have built up Caribbean reefs over the last few thousand years through the slow accumulation of their calcium carbonate skeletons. The gigantic structures they create are occupied by thousands of other species that couldn't survive without a refuge from predators like sharks and barracuda. Ecologists call these habitat-forming organisms 'foundation species'. Trees fill the same role in forests. When corals and trees die off, the loss cascades through the ecosystem, and countless other species subsequently disappear. Diverse, vibrant ecosystems are wholly dependent on dense, healthy populations of these and other foundation species.

The jarring reality is that Caribbean reefs have lost more than half of their reef-building corals over the last 25 years. On many of the reefs I study, nearly all of the coral is gone, and the seafloor is now covered by seaweeds, sponges, bacterial mats and other critters. Like tree die offs, the Caribbean's corals are being killed off primarily by diseases, many of which are caused or at least enhanced by ocean warming. Although there's a fairly clear association between high temperatures and coral disease outbreaks, the underlying mechanisms are largely unknown. The possibilities include a weakened coral immune system (due to ocean warming), newly introduced pathogens (e.g., from sewage or coastal deforestation), or mild winter temperatures that enable parthenogenic fungi and bacteria to overwinter. Some corals also simply can't stand the heat and are killed directly when water temperatures exceed their thermal tolerance (a phenomena called 'coral bleaching' by reef scientists).

Climate winners and losers

Some coral species are more sensitive to climate change than others. But unfortunately, like in the mountain forests I cycled through, the massive, long-lived foundational species seem to be much more sensitive than the small, weedy species they are being replaced by. The weedy plants and corals can't fill the same ecological role as the trees and massive corals that have dominated these ecosystems since before humans evolved.

While some species are clearly harmed by climate change, others can benefit from it. For some types of organisms (e.g., corals and trees) it appears to be the large, long-lived ones that are more susceptible. In some cases, it's because they're more physiologically sensitive to environmental change. Other times, species that are climate change 'winners' are equally sensitive but have a greater capacity to repopulate areas after a climate disturbance (like a storm or fire). This is the case for some weedy species that allocate a lot of energy into producing massive numbers of offspring able to travel great distances and repopulate denuded areas. They also benefit from the absence of the once-dominant climate change losers.

My first climate catastrophe

The first time I saw clear, large-scale impacts of climate change was in 1998. I was in Palau (a tiny island nation in the western, tropical Pacific) assisting my PhD co-advisor on a global biodiversity assessment. On our very first dive on Palau's famed coral reefs, we knew something was drastically wrong. Nearly all the corals, especially the dominant plating species in the genus Acropora, were pale white. Soon, most were dead. Within a week, their exposed calcareous skeletons were covered with a green-brown film, as they were colonised by micro-algae.

During the following weeks, we documented the impacts with video transects (via SCUBA). Our local collaborator, Dr. Pat Colin, assessed bleaching across the region in his homemade ultralight aircraft! (You can easily see bleaching from the air hundreds of feet above a reef.) Roughly half of the nearly one thousand coral colonies we surveyed across nine sites were bleached, and the die off was widespread across the main atoll and neighbouring reefs.

The cause was unusually high temperatures—about 1°C greater than normal for this region. By early 1999, reefs around the world had bleached and mass-coral mortality was being reported by hundreds of teams working in dozens of countries. It was the first global coral-bleaching event and arguably the largest, short-term impact humans have ever had on the natural world. And yet few people are even aware it happened.

Climate science fatigue

Only a tiny fraction of humans ever witness climate-induced degradation of ocean ecosystems (or read journal articles about it). It's almost never covered by the media and even nature documentaries generally downplay it. In contrast, nearly all of us are seeing and feeling changes to our local weather conditions. Every person I talked to along my trip—including cowboys in Montana and miners in Colorado—was aware of the changes in nature and the environment around them. You really can't live there and not notice the tree die offs.

Despite the devastation in Palau, I don't remember any of us being especially sad or otherwise emotionally affected by what we saw in the moment. At the time, we were in awe of Palau's staggering biodiversity and colourful invertebrates and we were extremely busy and largely focused on the work. I think we also had little inkling of what we were really witnessing. We didn't realise how common it would become, or the transformative impact ocean heating would have in the following years.

More recently, I've slowly become aware of how my work on climate affects me emotionally. It certainly gives me a strong sense of purpose and community, but it often makes me feel angry, disappointed and impatient for change. The topic has dominated my consciousness and conversations for over two decades. I'm feeling the climate fatigue and anxiety many of my colleagues are talking about. Still, I'm trying to be more mindful about how my work on climate and conservation affects my wellbeing (and indirectly affects the people around me). I've recently started meditating and nurturing the feeling of being grateful for the natural world.

Writing climate obituaries for the natural world

One substantial limitation of our Palau study was the absence of pre-impact data. As a result, we couldn't measure how the bleaching affected coral abundance and composition. Ecological monitoring isn't always exciting, but it's proved crucial to quantifying climate change impacts. Scientists working in all kinds of terrestrial and aquatic habitats revisit their field sites annually to track changes in species abundances. This usually means counting, photographing, or otherwise recording population densities in dozens or hundreds of plots scattered across multiple sites. All that data is then entered into spreadsheets, archived and often shared with other researchers.

After graduating from Brown with my PhD, I got involved in reef monitoring, first in Mexico, and later in Belize. At the time, I didn't expect much to change (at least in the short term). But within five years, we were already documenting mass die-offs of some of the most abundant and ecologically important coral species on Caribbean reefs. In 2000, I tagged several hundred colonies of the massive boulder coral *Orbicella* for a demographic study. By 2005, nearly all were dead. Until then, the mass-bleaching I'd experienced had been of fast-growing, short-lived, weedy species that could recover within a decade or so. But these *Orbicella* colonies—all at least a meter or two in diameter—were centuries old. These populations weren't just affected; they were nearly extirpated and would take many human lifetimes to recover.

The loss of these and other ancient organisms is probably what affects me the most. These impacts are nearly irreversible, a lot like clearcutting an old-growth forest. The skeletons of these massive corals are still there and probably will be for decades. Now they are overgrown by the weedy species that have come to dominate Caribbean coral reefs. For now, this is the state of a lot of the dying conifer forests I rode through. The standing dead trunks of the trees remain—at least for now, until they fall over or burn.

Aspen provides a wakeup call

My five-week bike trip ended in Aspen, Colorado. My daughter, Mazarine, age 16, joined me on the trail for the final few days. We rode across some of Colorado's highest peaks and through lush alpine valleys. We saw moose grazing next to clear mountain streams and countless meadows blanketed by wildflowers. We were chirped at by marmots and pikas in boulder fields, and on the last night we camped next to a shallow alpine lake at 12,400 feet, just below the final pass we'd cross the next morning. This was my last and favourite campsite of the trip. The high peaks of the snow-covered Rockies surrounded us. The ground was covered with outcrops of white quartz and ancient and hearty alpine plants taking advantage of the short growing-season. The sky was clear without any moon, so once it got dark the number of stars and the brightness of the Milky Way were bewildering.

But the most striking thing was the quiet. When there's no wind, the total absence of sound in the wilderness, especially at night, always surprises me and can be intoxicating.

Aspen is one of the world's richest and most exclusive communities. On the way into town, we rode by the airport and gawked at the dozens of private jets lined up on the runway. Mazzy and I talked about the stunning wealth and the irony that the emissions from these toys of the super-rich were destroying the places they ferried their passengers to. Mazzy and her generation are acutely aware of the climate crisis and who's responsible. We talked about the future a lot on the trail and although she has the climate anxiety of most teens, so far it doesn't seem to have darkened her outlook. She's optimistic about her future, which is becoming a rare feeling for teens worldwide.

She told me that having spent most of her life in nature, she can feel helpless, especially knowing she's a kid who doesn't have much power over the fate of our planet. 'I worry that by the time I'm old enough to have a professional influence it will be too late,' she explained as we cycled together.

Mazzy is a realist, but she's also adept at seeing and getting excited about the smaller things in the here and now. The bees we saw pollinating wildflowers in mountainside meadows. The pikas chirping at us. The bark of the old, gnarled pines you see at high elevation.

'Seeing the changes caused by climate change firsthand could easily lead to a gloomy mindset,' she told me at one point. 'But I know how important it is to reflect on the beauty, complexity, and importance of the nature that I am experiencing as a call to protect it and to never stop fighting for the greater power that is the natural world. I do try to hold onto the hope.'

So far, she has held onto that ability to observe, and the feeling of wonder, that most kids have. And I'm working on doing more of it myself (and worrying less about climate models and political obtuseness).

Although the sky was crystal clear the last week of trip, when we flew out the next day, smoke had enveloped the area. We couldn't see the hillsides or mountains adjacent to the airport and our takeoff was delayed due to poor visibility. All of this was an abrupt reminder that these climate-fuelled conditions are real, they're getting worse, and we still don't truly understand how big a threat this is to our homes, our communities, our economy, our health and nearly every aspect of our lives.

What to do? Eating less (or no) beef, flying and driving less and riding our bikes more can all help. But we also need to begin a structural transformation of the nation—of our transportation and energy systems, of our housing economy and healthcare system, of our management and protection of ecosystems and, most of all, of our political systems. We need to start now and reduce our greenhouse gas emissions dramatically in the coming years and decades. In truth, we needed to start decades ago.

Will we do all this and limit the warming to around two degrees Celsius? (So far, we've warmed the planet by around one degree Celsius.) I have no idea how to assess our chances, but honestly, I wouldn't bet on it. I'm going to continue to do what I can to increase our odds, but I'm also going to be doing a lot more bike-packing and exploring of our world. Despite climate change and countless other environmental problems, there's still so much wonder and natural beauty out there. There's so much left to marvel at and to protect. And it's good to be reminded of that.

SENTINELS OF THE ECOSYSTEM

BY

KATIE HOLT, WITH ASSISTANCE
FROM DR DEE BOERSMA

Katie Holt *is a PhD candidate at the University of Washington, Seattle, under the study of Dr Dee Boersma. Her work focuses on how species, particularly seabirds, can be early warning signs of changing environments. Species that are sensitive to environmental change and easily observable can be informative ecosystem sentinels by indicating otherwise unobservable change in the environment or ecosystem. Seabirds are often used as ecosystem sentinels because they are easily observed on land but forage exclusively at sea. This allows scientists to learn about oceanographic conditions, like prey availability, that are difficult to quantify but can affect entire ecosystems.*

sat cross-legged in the dirt, drenched in the smell of death. I was familiar with this smell. I had been working in this Magellanic penguin colony in Patagonia for the past two years. Occasionally, this distinct smell would get caught on a breeze and I would follow my nose to find the dead penguin and investigate why or how it died. Death was typically relatively rare, so the smell usually did not bother me. My curiosity would kick in and I would want to know what had happened.

This time, the whole colony reeked of death and dozens of vultures circled overhead. My clothes and research gear were covered in goop from measuring hundreds of dead penguins. I didn't have time or sufficient water to wash my clothes frequently because water had to be brought 120 kilometres from the nearest town. All my measuring gear—callipers, rulers, weigh scales-was caked with this dead goop. I was devastated by the extent of the mortality. But I also saw this as an opportunity to learn—I knew a single-day mortality event of this magnitude had never been seen in the 40 years my PhD advisor, Dr Dee Boersma, and her students had been studying this colony. I was determined to collect as much information as I could before the penguins decomposed too much or were eaten by scavengers.

The study started because a Japanese company wanted to kill penguins for oil and use their thick skins to make golf gloves. Fortunately, no penguins were ever killed at Punta Tombo and, instead, tourists come to see and enjoy the expansive colony. Dee's long-term studies, which have spanned topics such as how climate change affects penguins and why penguins get divorced, have led to effective conservation policies.

I started working with Dee in 2016 as a lab manager, running lab logistics and helping with fieldwork. In my first two years, I spent 24 weeks at Punta Tombo assisting with field studies. This was my third year helping with fieldwork and my first year as a graduate student. This season I was joined by an undergraduate volunteer, Anna, and a newly hired lab manager, Maria. Anna was a tall, fit rock climber, interested in studying oceanography. She was eager to help with all tasks. Maria was a retired lawyer with impeccable attention to detail, looking for a stimulating new job.

Punta Tombo is one of the largest breeding colonies for Magellanic penguins, with around 120,000 breeding pairs. From Dee's long-term work at the colony, we know it decreased by 36 percent from 1987 to 2014 and continues to decrease. The colony sprawls across a coastal, arid desert in windy Patagonia, where penguins nest in bushes and burrows for protection from the elements and predators. Some penguins nest over one kilometre inland. At first glance, the landscape appears desolate. To me, and likely others who are familiar with Patagonia, it thrums with wildlife in one of the most wild and untouched places in the world. Here you can find the elusive puma, giant herds of guanaco, bouncing mara, the dinosaur-like rhea, thousands of sea lions and a few dozen elephant seals lounging and snorting on the beaches, and of course, during the breeding season, the near constant bray of hundreds of thousands of Magellanic penguins.

19 January 2019: the hot day
Several days before I was covered in dead goop, Maria, Anna and I were making the morning rounds checking marked nests. Since the project began in 1982, Dee and project volunteers have banded around 60,000 penguins at Punta Tombo to learn what threats they face and to suggest conservation actions to local management. All nests where we know or suspect a banded penguin will inhabit the nest get marked with a bright yellow plastic tag—the same tags used by ranchers to identify cattle. Our system for finding and keeping track of all our marked nests consists of distance and direction from one nest to another (e.g. 25 metres NNW to nest H07Y).

This morning, we checked nests in a section of the colony close to the ocean. We started work early, around 6:30 or 7am. In Argentina, January is the austral summer, so we start days early to avoid working and handling penguins in the heat of the day. We usually stop around 11am, take a siesta, cook a big meal, and return to work in the afternoon when air temperatures are cooler. The long days in summer allow us to avoid the heat of the day and still have daylight to work long hours.

I started to feel uncomfortably hot around 9:30am. This was strange because we had a clear view of the ocean, which usually afforded a cool ocean breeze. However, I do not trust my internal temperature gauge. Having grown up in the Pacific Northwest, I am what some would call 'Seattle soft' and I don't handle extreme heat or cold well. I felt I should put on a brave face and encourage Maria and Anna to keep working as well. However, by 10:30am we had all run out of water and felt lightheaded. We headed back to the house early to rehydrate. With the extra siesta time, I decided to set up my hammock between beams on our front porch and take a nap. When I woke an hour later, I felt horrendous. Dry, hot wind was blowing in from the west, leaving me very dehydrated.

In the afternoon, we set out again to continue nest checks. Running through a section of the penguin colony is a gravel trail for tourists. As we walked out, several tourists stopped us to ask why there were dead penguins. Thinking they had seen one or two dead penguins from the trail, we explained about 39 percent of chicks die of starvation and that some chicks and adults are predated by foxes, puma, giant petrels and weasels. After completing our evening nest checks we found three dead adults and a dead chick on our way back to the house. I assumed these individuals died from heat stress. It had been a hot day and they were all found dead in a position that indicated they died trying to release excess heat—flippers and feet extended away from the body and bill open like they had died panting.

The sun was setting, so we quickly performed necropsies to rule out other causes of death. We examined all individuals and found no apparent signs of predation before we cut into the belly to check the stomach lining for signs the penguin ate prey contaminated with toxic algae. If the individual died from toxic algae the stomach lining would be a bright red or pink colour, and the irritated stomach would produce lots of mucus. The stomach lining looked healthy, further solidifying the individuals died from heat. In previous years at Punta Tombo, I had seen a handful of penguins that had likely died from heat. These penguins had thick layers of fat underneath the skin and were likely getting ready to moult. Penguins go through a 'catastrophic moult', meaning they shed all their feathers at once and need to fast on land for an extended period until all their feathers are changed. I noted the adults we found that evening had very little or no fat under the skin, which seemed unusual.

Usually, we wait until the following morning to record the maximum temperature for the day, but out of curiosity, we decided to check on our way back to the house. The thermometer read 44 degrees Celsius. For those working in the imperial system that is 110 Fahrenheit. The thermometer was in a large, shaded bush out of direct sunlight, meaning exposed sections of the colony were likely even hotter. Over dinner, we checked our database spanning back to 1982 and found it was the highest temperature we had ever recorded in the shade at Punta Tombo. However, there were a couple of days with a maximum temperature of 43 degrees Celsius in the 1980s, so I didn't think it was anything too extraordinary (even though the average maximum temperature during the summer is only 26 degrees Celsius, plus or minus five degrees).

20 January 2019

The next day we searched for banded penguins in a central section of the colony on top of a hill with an expansive view of the ocean. This section is farther inland, so there is no ocean breeze. I always enjoy searching this area because this is where many penguins were banded a couple of decades ago. The density of this area has dramatically decreased over time, and the easiest way to find penguins here is to hear them 'sneeze'—blowing the excess saltwater out of their nares (nostrils) from drinking seawater.

When you do find a banded penguin in this section, it is likely to be at least 20 years old. Some I found that day were older than me at 26 years old. Because this long-term study has banded penguins since 1982, and followed individuals over time, we know that a few penguins banded as chicks have lived into their 30s.

As we checked nests for these older, banded penguins, we started finding more and more dead penguins. All dead adults and chicks were found in a similar position—flippers and feet extended from the body to release heat and/or bill open, indicating they died panting. All were freshly dead. We performed necropsies and found many of the chicks had huge bellies full of food and there was no evidence of other causes of death like disease, toxic algae, or starvation. Because necropsies are time-consuming, and we were finding so many dead penguins, we started keeping a tally and measured the bills of adults to determine their sex. We found 39 dead penguins that morning.

In the evening, we conducted a beach survey where we counted the number of adults and juveniles (one-year-olds) on the beach. This helps us identify if many chicks from the previous breeding season survived winter migration and returned to the breeding colony. We found 12 dead adults just metres from the water and one dead giant petrel.

By the end of the day, the magnitude of mortality from the hot day was sinking in. It was astonishing to see penguins dead only

SENTINELS OF THE ECOSYSTEM

a few metres from the sea. If they had been able to walk 10 or 20 more metres down the beach, they would have been able to cool off, drink ocean water and survive the hot day. On our way back to the house we chatted with the local rangers and discovered a handful of tourists had passed out on the tourist trail on January 19 and required medical attention. The tourist trail is only one or two kilometres long.

As a first-year graduate student with relatively limited experience in the field, I was starting to feel the pressure of ensuring I captured as much information about the heat deaths as possible. In previous years I followed a highly regimented protocol and never had to make too many difficult decisions on what data to collect. Our extensive protocol had instructions for some extreme events, like large rainstorms. In 2014, Dr Ginger Rebstock and Dee published a decades-long study showing how climate change has increased the frequency and severity of rainstorms near Punta Tombo, resulting in an increasing number of chicks dying in rainstorms. Unfortunately, our protocol did not include what information to collect when many penguins die from heat.

With limited internet contact with Dee and other colleagues in Seattle, I was intimidated by making decisions on what data to collect. I knew we had limited time to collect useful information. Each day that passed meant carcasses were decomposing or being eaten by scavengers. Maria, Anna and I spent the evening discussing our plan of action. We were physically and emotionally exhausted. We were walking around 10 miles a day searching about 240 hectares of penguin breeding habitat looking for dead penguins. Our rapid necropsies took at least 10 to 20 minutes as we weighed, measured, and assessed each dead penguin.

21 January 2019

We decided to measure 30 random chicks that survived the hot day in an area of the colony where we had found the most dead chicks. This would allow us to compare the body condition of chicks that died to chicks that survived. Body condition is like the body mass index (BMI) used with humans. If a penguin has a high body condition index, then it is relatively fat for its size and vice versa. I initially hypothesised adults and chicks that had higher body condition index (relatively fatter for their size) would be more likely to die in the heat than individuals with a lower body condition index because in the past I had found adults dead from heat with thick layers of fat under their skin. Instead, we found the opposite for adults. Dead adults were in significantly worse body condition than adults that survived.

Adults get all their hydration from their food or drinking seawater (they filter excess seawater out with the supraorbital gland above their eyes). When they are tucked in their nests in the colony, they cannot access food or water. These penguins are used to fasting on land as they incubate eggs for a couple of weeks; however, incubation is during a cooler time of the year. We predict the adults that died on the hot day had been fasting on land and were too dehydrated to pant through the hottest part of the day.

The body condition of chicks that died in the heat and chicks that survived were similar. However, we did notice chicks that died had huge bellies full of food. It seems the large meal from their parent made it more difficult for the chicks to thermoregulate through the hot day because blood would have been needed for digestion and less available for cooling. In contrast to adults, chicks would have been well hydrated as they get all their water from their meals. We had found 80 dead penguins in the past three days.

22 January 2019

It was much cooler on the 22nd. We even wore jackets! We weren't continuing to find any dead penguins, so we started to relax, thinking the emotionally and physically laborious work of finding and documenting dead penguins was over. We returned to our scheduled data collection and protocol. One of those tasks was catching chicks fledging (leaving the nest and heading to the ocean for the first time) to weigh and measure them. Weighing chicks as they head to sea for the first time tells us whether adults were finding enough food to feed their chicks to survive their first winter at sea.

23 January 2019

With our limited internet connection, our colleagues in the lab warned us the next day was predicted to be another hot day at 39 degrees Celsius. The atmosphere of the camp was relatively glum because we were dreading another wave of dead penguins.

24 January 2019

Relief set in slowly throughout the day as we searched the main study areas of the colony and only found two new dead adults. We were worried adults would still be stressed from the hot day on the 19th and one more hot day would lead to more mortality. Some studies have documented mortality from extended heat stress, while others found heat death events occur because an upper-temperature threshold is crossed. Because we only found penguins that appeared to have died on the 19th (the dead adults all had a similar amount of bloating) and temperatures were relatively mild in the days prior, this event seems like an upper-temperature threshold was crossed at a time when penguins were vulnerable because they were dehydrated after fasting on land.

25 January 2019

In the morning, we set out to survey a southern beach section of the colony. We left the house feeling a little lighter, thinking we wouldn't run across many more dead birds. The southern beach is a section of the colony that is slightly outside our main study area, so we hadn't visited this section of the colony since the hot day. There are several 'penguin highways' here that penguins use to enter and exit the nesting area from the ocean. On each highway we passed, we found a half dozen dead adults that appeared to have collapsed leaving the colony.

The scene at each highway was gruesome. Some adults had their bills stuck directly into the ground, or their heads smashed on a rock, indicating they had abruptly collapsed attempting to flee the heat. Again, we started trying to necropsy and measure all the dead penguins, but we realised we would be working all day if we proceeded at that speed. On the beach we found 43 dead adults, metres from the waterline and scattered along the berm on their way to the ocean. Again, it was shocking these birds were such a short walk from the water where they would have been able to cool off and rehydrate.

We found 264 dead adults and 90 dead chicks that died in the heat on January 19, 2019. More males died than females, but this is unsurprising because work by Dr Natasha Gownaris has shown that males are becoming more common in the colony because females have a higher mortality rate than males. Currently, we estimate there are three males for every female. Mortality on the hot day was unevenly distributed across the colony. More penguins died in inland sections of the colony compared to sections close to the beach, indicating some areas may have been a refuge from the heat or had easier access to the ocean than others.

The long lifespan of penguins and other seabirds allows populations to persist through years when many chicks die. In some years, most chicks die of starvation or because of extreme weather events like rainstorms, but the long lifespan of these penguins means adults will likely have many years to breed. Thus, adult mortality rate has the most significant effect on population growth rates. Rare and punctuated mortality events, like the heat mortality at Punta Tombo, have the potential to have severe consequences for the survival of the colony.

To provide some context for the magnitude of the heat mortality, we examined how many of our banded adults were found dead from heat since the study began. Between 1982 and 2019, the project banded 16,738 adults and later found 391 of those adults dead. Of those dead adults, only seven appeared to have died from heat. Because heat deaths were so rare, a mass mortality event like this was very unexpected. In seven prior years, temperatures reached between 40 and 43 degrees Celsius, but we did not observe any mass mortality.

Even if penguins do not die in the heat, heat avoidance behaviour could have negative consequences. On hot days, adults are more likely to spend time on the beach, which means breeding adults would leave eggs or chicks in the nest alone, making them more vulnerable to predators. Unattended eggs are quickly snatched up by gulls or armadillos.

Much of climate change research focuses on how or if species will adapt to the changing environment. One way species persist through changing environments is by shifting where they live to environmental conditions they prefer. For example, colonies north of Punta Tombo are growing, likely because they are closer to their food and ecological conditions, like temperature, are better. San Lorenzo, a colony 240 kilometres northeast of Punta Tombo, recently surpassed Punta Tombo in size. Penguins from Punta Tombo could be moving north to San Lorenzo. It is unlikely the penguins at Punta Tombo could adapt to survive through a hot day because the hot days are so rare, but penguins, especially young penguins just starting to breed, could be choosing to move to other colonies.

With global climate change causing an increase in the frequency and severity of extreme weather events, heat-related mortality is expected to increase for many species, humans included. Many people, especially people in wealthy countries, have historically been buffered from the direct effects of climate change. The unprecedented heat mortality in a single day was the most tangible evidence of the devastating effects of climate change Dee and I have witnessed. These penguins are ecosystem sentinels, and they are telling us it is getting too hot.

REVISITING THE NUNS VEIL

BY ——

HEATHER PURDIE

Dr Heather Purdie *is a field-based glaciologist who has been researching and teaching students about snow, ice, glaciers and climate change for more than 15 years. As an Associate Professor at the University of Canterbury / Te Whare Wananga o Waitaha, she enjoys getting students out into the field so they can experience the environments they are studying and learn about processes first-hand. Heather has a number of ongoing glacier monitoring projects in the New Zealand Southern Alps and works closely with local communities whose livelihoods revolve around snow and ice.*

There is a distinct satisfaction in listening to the rhythmic sound of your ice axe and crampons biting into a frozen snow and ice surface. Jason and I could not see much that morning. We had begun in the dark, the colder predawn temperature reducing the risk of rockfall. As the sun rose, the thick fog was almost a whiteout, and it engulfed the Nuns Veil in cloud as we systematically worked our way up the glacier towards the summit.

We were moving un-roped as it is not a difficult climb—a Grade 1+ for those who mountaineer—and one I had done six years earlier in 2007. The route follows the Nuns Veil Glacier up the southern flank of the mountain, rising in a series of gentle steps. Occasionally gaps would form in the swirling mist, and we would get a glimpse of the summit.

My passion for mountains and glaciers was instilled at an early age. When I was seven years old, on one of our few family holidays, we went to Aoraki Mount Cook National Park. I remember long walks with my family and crossing the Hooker swing bridge. We also went to view the *Haupapa*/Tasman Glacier. There was no lake in front of the Tasman Glacier then, and it was blanketed in rock, though you could still see bits of ice poking out. I could hear the rocks and ice creaking and cracking as the glacier slowly flowed down from the distant mountains.

On that trip, I learnt that during the ice ages these amazing things called glaciers used to be much bigger, and that concept fascinated me. I imagined how it would have looked when the ice where I stood was so thick that only the tops of the mountains poked through.

That early family holiday was followed by studying geography at high school, which included a field trip back to the Tasman Glacier. At the same time, I developed my own recreational pursuits that involved climbing in the hills and mountains. As a teenager, I headed off across the Tasman Glacier with a friend, intending to hunt chamois and tahr in the Murchison Valley. Both are introduced species with no predators and, in high numbers, can damage alpine vegetation, so we hunted them for meat and trophy heads.

The moraine, a thick covering of jumbled rocks on the ice surface, made walking with our large packs slow going. Our progress was further impeded by a number of small melt ponds we encountered, which required circumnavigating. One particular melt pond was already much larger than the others; although only 50 metres wide, it was long enough that it took us at least an hour to detour around it.

My friend and I had been given directions on where to cross the glacier by people who had been there in recent years, but none of them had mentioned the lake. We looked at each other when we saw it and wondered, 'What does this mean?'

It was the 1980s and climate change was not really a big thing at the time. I was aware of it mainly because I was a geography student, but it wasn't something people talked about or thought about very much. Seeing that lake was the first time I had encountered something I knew was a direct result of rising temperatures and I realised it was what I was learning about. I thought about just how big that pond was going to get and what it might mean for future access.

Who knew then that 30 years later I would be standing on the same moraine explaining to a group of university students why the area I had once walked on, and where my sister and I had sat as children, was now a lake more than six kilometres long. These days, it's getting difficult for the students to even see the glacier's terminus in the distance at the far end of the lake.

It's not just the Tasman Glacier that is shrinking, of course. Today all around the world, glaciers are melting fast because the once natural oscillations of climate have been disrupted by human-induced warming. Glaciers gain and lose mass naturally as the climate varies, which is why they are such excellent indicators of climate change. They exist on the surface of the Earth in a very delicate balance; if temperatures increase, more snow and ice will melt, and if the climate gets colder, more rain will fall as snow. The *Te Moeka o Tuwae*/Fox Glacier, for example, is now 3.5 kilometres shorter than it was in the 1800s when Europeans first started exploring there—shorter than any other time in recorded history. It has been retreating at a rate of approximately 90 metres per year since it last advanced in the late 2000s.

Now that the Tasman Glacier ends in a lake, it behaves differently to glaciers that terminate on land, and is subject to different processes and feedbacks. Once there is a large body of water lapping up against the end of a glacier, the drivers of ice retreat and the rate at which that retreat occurs, change. The water is always warmer than the ice, so as soon as you have the water in contact with the glacier, the water melts the ice. Warming water along with warming air means the glacier is under attack from all sides.

Ice ramps may also influence the speed that the glacier retreats. At the terminus of these glaciers, where the glacier meets the water, underwater ice ramps or shelves of ice can form. These influence the buoyancy of the glacier. It's a complex process—and one that I'm currently researching—but when the glacier becomes thinner, the end can start to float in the lake water. This puts the ice at the end of the glacier under a lot of pressure, resulting in large iceberg calving events.

This is important not only because the underwater ramps influence the rate that the glacier is shrinking and the lake is expanding, but they also present a hazard for people using the lake. These underwater ramps are hidden from view because glacial lakes are extremely silty. When the hidden ramps calve or break off, they create huge waves and fill the lake with ice.

I work closely with the local guiding company, Glacier Explorers, and it's apparent that they, as anyone working or exploring in snow and ice, are now operating in a rapidly changing environment. People in these environments must be aware that the rate the ice is changing is so fast that you must be dynamic and learn to adjust based on year-to-year changes.

As Jason and I continued up the Nuns Veil, we reached the bergshrund (the crevasse separating the moving ice from the ice on the upper mountain). It can sometimes be wide open and tricky to cross, but on this trip it presented no problem as it was still full of snow from the winter. We crossed and started up the final slope towards the summit. I had hoped the mist might clear as we got higher, but not that day. I was just starting to ponder whether or not the top portion of the slope seemed a little steeper than last time—I didn't remember the climb being this hard—when with a 'twang' my ice axe punched through the ice and bounced on the rock below.

I had a definitive answer to my question now—the ice cloaking the mountain was definitely thinner this time around, and the route was indeed becoming steeper. Thinning ice can make a climb more difficult over time. While I was experienced enough to navigate the climb, it was a reminder that the mountains are becoming less predictable. These days, my research into glaciers and the changing alpine environment focuses largely on glacier mass balance, including the feedbacks or flow-on effects of climate change on glacier health. And the ability of my ice axe to penetrate right through to the underlying rock was a clear sign that the glacier was losing mass.

My research comes very much at the intersection of work and play. For me, the mountains are not just about work; they are where I grew up. Climbing brings a sense of connectedness to the environment; you become very tuned in to the weather and the snow conditions, you feel the crispness of the air and the screech of the kea.

So when I started to hear talk in the mountaineering community about climbing routes becoming 'cut-off' as glaciers 'break up earlier' or becoming 'more convoluted' as glaciers respond to climate warming, I wanted to investigate. Such anecdotes, coupled with my own experiences and observations, inspired myself and a friend and colleague, Tim, to explore what changes could be detected on the main climbing route on Aoraki (Mount Cook), the Linda Glacier route. Tim and I had both climbed the route on separate occasions in the past and both enjoyed delving into photographs and archives of early mountaineering history.

We combined measured physical data with conversations with guides and recreational climbers. We found that at lower elevations, glacier retreat was clearly evident and impacting people's access to the climb. The thinning or 'downwasting' of the Tasman Glacier has resulted in very large moraine walls, which are steep unstable slopes that can be difficult to descend, making foot access to the route more time-consuming and challenging.

High up the mountain, climate change was not so obvious and measurable. However, climbers talked about their experiences with crevasse exposure, slope steepening and rock exposure. Some people noted that rock was exposed in places that 'didn't exist on early summits'.

The large year-to-year variability in snowfall received in the New Zealand Southern Alps makes it hard to quantify changes in snow frequency and volume over time. But after completing the Aoraki climbing project, the conversations I had with climbers still lingered; there was much left unanswered. These unanswered questions formed the foundation of my latest research programme. When the crevasses are exposed at the surface they increase surface roughness and trap solar radiation; my research hypothesis is that the models we currently use to estimate melt rates on glaciers will be underestimating snow melt in the accumulation area.

So, for the last two summers, I have been leading a team at the top of the Tasman Glacier where we install temperature sensors inside crevasses and use drones to map the surface of the glacier with thermal imaging. Crevasses are formed as the ice moves. In the winter they are usually covered with snow, then as the summer progresses the snow melts and they are exposed at the surface.

My first field season was in 2020. I knew what I wanted to measure, but exactly how our bespoke 'crevasse temperature sampler' was going to perform was a great unknown. I had hand-picked a research team that I believed had the right skills and experience to make things work: Jane is an experienced mountain guide; field technician Paul and post-graduate student Ben are both on the cutting edge of working with new technologies; and my friend and fellow scientist Tim had worked with me on the Aoraki project.

One particular challenge of working with new technologies in remote environments is power. Thermal cameras, drones and weather stations all use power and generally lots of it, so not only is your team roped together for glacier travel, but each person is also juggling tripods, solar panels and big batteries; we probably look quite a sight. Our bespoke crevasse sampler consisted of 16 metres of PVC pipe, 20 temperature sensors, two 'beta' wind speed sensors and lengths of rope to hopefully lower the unit into a crevasse without damaging it.

The first task for the team was to locate a crevasse that was deep, open and had wide enough spacing between it and other crevasses to give us room to anchor our equipment and work safely. The sampler looked unwieldy lying along the edge of the crevasse and there was a bit of deliberation and contemplation about exactly how we would coordinate the operation of the rope pulley system designed to help lower the unit. But in the end teamwork prevailed and we got the unit in place and gathering data.

There is still a lot to do, but our preliminary results show that at times crevasses can be surprisingly warm. Previous work on air temperature in crevasses was only done in polar environments, so there is limited information on air temperatures in crevasses in maritime environments such as New Zealand. What we have found so far is that at times the air in a crevasse is even warmer than the air on the surface of the glacier, meaning melt could occur even faster than previously thought.

Beyond this, for my broader research, I spend a lot of time in the New Zealand Southern Alps, studying the Tasman, Fox and

Rolleston Glaciers. This includes monitoring the mass balance of the glaciers, which is essentially a glacier health check. I measure how much snow the glacier is accumulating each year versus how much snow and ice is melting away. The mass balance is the amount of mass (snow and ice) left on the glacier at the end of each year. If a glacier is gaining more mass than it is losing, then it will grow bigger, getting thicker and longer. Conversely, if the snow and ice melt exceeds snow gain, then the glacier will get thinner and shorter.

My work on the Rolleston Glacier near Arthurs Pass focuses on longer-term monitoring. A team of us visit the glacier twice each year: once at the end of the winter to measure how much snow has accumulated on the glacier; and then at the end of the following summer to measure how much winter snow is left and how much the ice surface has melted. By combining these data we can calculate the mass balance of the glacier. On these trips we dig snow pits, probe snow depth and drill stakes into the glacier to measure ice melt.

Although advances in satellite technology mean we can now do quite a bit of glacier monitoring remotely, it is still crucial to have sites where people actually go into the field and measure what is there. Partly, this ensures we are interpreting the satellite data correctly. Also, while satellites can be used to measure glacier area, they are unable to provide information about snow density, which enables us to work out how much water is being stored in the mountains, a characteristic that is very important for water management.

The Southern Alps of New Zealand typically get a lot of snow due to their maritime location, and that snowfall tends to be highly variable from year to year. So, over the last decade, we have not recorded significant changes in the winter balance of Rolleston Glacier. However, we are starting to see summers when the melting is much more extreme. In 2018 a team of us arrived at the glacier only to find that nearly the whole glacier surface was ice. Almost none of the snow that had fallen the previous winter had survived the summer heat. It was a very sorry sight. The ice was dirty and we could not dig a snow pit because there was no snow left to dig; all we could do was pick up the stakes that had melted out of the ice. We faced the same situation again in the summer of 2019.

When we began the project 10 years ago, we used to climb up the slope to the saddle and step straight onto the glacier. Now there's a gap and at the end of summer we need to scramble down the rock and climb back up onto the glacier. The gap isn't huge yet, but it's noticeable that the glacier is shrinking and pulling away from the wall. It gives me a real sense of loss to watch the demise of what used to be a defining feature in the alpine landscape. Instead of the vibrant, healthy glaciers I used to see, I've watched them become diminished and dirty, shrivelling away.

One thing that strikes me in my work on climate change is that people are more likely to care about it if they can feel and experience a real connection to the environment—it makes it more real. It is this sense of connection to the environment that I try to instil in my students. After all, they are the ones inheriting Earth, so we need them to care, to learn how it works and try to find some solutions to the mess humans are making.

But getting people to care about change can be challenging, especially when the glacier they have come to visit is no longer accessible by foot or is not the majestic glistening feature they expected: rather something that has transformed into a small, dirty mass of ice, viewable only from a distance or from a helicopter. On the Tasman Glacier it can be challenging for visitors these days to actually see the thickness of the ice and, because it's so far away and covered in debris, visitors can be a bit underwhelmed.

As highlighted on our climb of the Nuns Veil, when glaciers are receding, they are not only getting shorter, but they are also getting thinner. This thinning of the ice mass contributes to several flow-on effects beyond the glacier just getting shorter. In particular, as the surface of the ice thins, it exposes rock at the valley sides. This rock is often loose and crumbly and cascades down the slope onto the ice surface below.

Rockfall is clearly a hazard for those walking on the glacier. Anywhere where the ice mass is thinning and disappearing, it will be exposing rock underneath, creating new hazards for anyone in these environments. This is exacerbated by the fact that freezing and thawing of ice breaks apart the rocks, so they are already very fractured.

This is not limited to New Zealand, of course. It happens around the world and in the European Alps increases in rockfall in alpine environments has also been observed and recorded as the ice recedes. So monitoring the rate at which ice masses are thinning is also important for understanding how such hazards evolve over time.

As well as this, as the ice melts, fragments of rock, eroded and transported down the valley by the glacier, are left behind on the ice surface, adding to the volume of rock that also accumulates from the valley sides. A very thin layer of debris on the ice surface can speed up the rate at which the ice melts, as the darker coloured debris absorbs more heat from the sun and transfers that heat into the ice.

However, once that accumulation of rock starts to exceed two to three centimetres in thickness, the rock begins acting as a protective cover, slowing down the rate at which the ice melts. It's hard to say how effectively and how long this could preserve a glacier. It might mean that for low angled glaciers, like the Tasman, a section of ice could survive in the bottom of the valley, disconnected from the mountain glacier that once fed it. Relic ice like this could potentially persist for decades after the active glacier has detached and retreated high into the mountains.

From a global water perspective, melting all of New Zealand's glaciers would not change sea level very much, because despite there being 2,900 glaciers in the country, the total volume of ice is small compared to locations like Antarctica or Alaska. However, glaciers in the Southern Alps are great natural laboratories because due to New Zealand's maritime location the glaciers here are very sensitive to climate change. They exist in a marginal environment, very close to freezing point, which means that even a small shift in temperature can have a big impact on glacier mass balance.

Globally though, glaciers are important not just for the volume of water they store, but the way they deliver that water into rivers. Glaciers at the top of a catchment store the rain that falls as snow during winter and then release it at the hottest time of year when water is most needed. This is not such a big deal in New Zealand as we have a lot of water. But in other parts of the world particularly in west coast South America and the southern Himalayas, communities and crops are entirely dependent on melting glaciers to supply water during the summers.

Earth is not just the home of people; it is also the home of trillions of plants and animals and is composed of an incredible array of landscape features. If we want the snow and ice to remain on the mountains, feeding the rivers that sustain us, Earth has to come first.

When our crevasse-team returned from the field work on Tasman Glacier in early 2020, the world had changed. Countries including New Zealand were heading into lockdowns in response to the global pandemic. As people adjusted how they went about daily life, global green-house gas emissions plummeted; so it can be done, we can live more lightly on Earth.

As I stood on the summit of the Nuns Veil the second time around, maybe it was a good thing Jason and I were engulfed in cloud and not able to see the ever-increasing evidence of climate change in the surrounding landscape; the lakes, the newly exposed rock faces and the diminishing snow and ice. What will it all look like if I return to the Nuns Veil again in future? Will I be climbing on snow and ice or will I be scrambling over loose, broken rock?

WHY MIST MATTERS

BY

AIDA CUNI-SANCHEZ

Dr Aida Cuni-Sanchez is an Assistant Professor at the Norwegian University of Life Sciences in Norway and an honorary fellow at the University of York in the UK. She has a PhD in Environmental Sciences from the University of Southampton (UK) and a Licenciatura in Biology from the University of Barcelona (Spain). She has over 10 years of work experience in 12 countries in tropical Africa, where she has focused on tropical forest ecology and carbon stocks, forest use and valuation by local communities and local communities' adaptation to climate change. She received the 2020 L'Oreal-UNESCO Women in Science UK Award for Sustainable Development and the Chr. Michelsen prize for outstanding development research (Norway) in March 2022.

'The problem is not just that rains have changed, it is also that mist has changed,' said the old man. And the other old men nodded.

Many other things have changed for the once-nomadic Samburu pastoralists of northern Kenya over the past few decades too. Because of so-called 'modernisation' and other political and economic pressures, they have been forced to settle down and live in permanent villages, reduce the size of their herds, divide their communal lands and start farming, send their children to school and pay taxes. But my research, and the day's discussions, were focused more narrowly on climatic changes, especially the increased incidence of severe droughts.

The old man was talking about mist though, not rain. This was going to be a long interview. Luckily, we were sitting under the shade of a large tree, and I had brought my water bottle. Our shady tree by the dusty road was the main gathering point—like a market square—for the village of Kituruni, just a 15-minute walk south of the Mt Marsabit mountain forest. Like most Samburu villages, small roundish mud-houses with circular fences made of thorny bushes to keep their animals safe from predators at night were scattered across the village.

'So, why does it matter that mist has changed?' I asked.

A heated discussion erupted which, unfortunately, I could only follow here and there thanks to the few sentences translated by Mohamed, my field assistant. I got the point though. Mist matters and it matters a lot. Mist (or fog) consists of millions of tiny water droplets suspended in the air near the ground and they are vital for many things. They help seeds germinate before the rainy season arrives, they keep the grass green and fresh for weeks after the rainy season has finished and people even collect the mist water when it drips from certain large trees. In short, mist is very useful to the Samburu people. Or rather, it used to be. According to the old men, the mist that used to regularly envelop the forest of Mt Marsabit had become very rare these days.

Mohamed attended a funeral over the next three days, so I was left on my own, trying to digest everything I had learned from the Samburu pastoralists. And the key word in all the conversations had been 'mist'. I wondered why the mist was disappearing. Was there a scientific explanation for this? The next day there was no power cut and I managed to learn more about mist in the small cybercafé in town.

While I am a botanist and ecologist by training, my expertise until then was the trees and rainforests of West Africa. In fact, I was only in Kenya as a favour to a colleague who asked me to fill in when the intended participant had to pull out at the last minute after breaking his leg. I had little knowledge of the trees of East Africa but had guessed that at least I could identify them because of my training in botany. So I had agreed, hesitantly. But now I was here, I enjoyed listening to the local people, particularly the elders, who were like walking dictionaries. They know so much about the environment where they live and the changes that have occurred—they pay close attention to the environment, as their livelihoods depend on it. They reminded me of my granddad, who I used to help collect wild edible mushrooms and berries in the forests of Spain when I was a child. Like the pastoralists, he knew all about the forest, even if he had never gone to school.

We all know that because of more carbon dioxide and other greenhouse gases in the atmosphere, temperatures across the globe have increased. Increased temperatures mean clouds form at higher elevations, so for mountains that reach into the clouds, the mist (which is really just a cloud touching the ground) now occurs at higher elevations—and lower mountains can be deprived of once common mist altogether.

This happens because air can only hold a certain amount of water vapour (water molecules in gas form), and this depends on the temperature (and atmospheric pressure) in a given area. The higher the temperature, the more water vapour the air can hold before it becomes saturated. Once it is saturated, condensation occurs, which is the process where excess water molecules in gas form change into liquid form. A large accumulation of such tiny water droplets in liquid form is a cloud—or mist.

So, there was a scientific explanation to the disappearing mist. And sadly, it suggested that things will only get worse with rising global temperatures. As I read articles and research papers online at the cybercafé, I realised that scientists have long identified the problem of 'raising cloud bases' related to global warming. Several studies from the Andes, Mexico and Costa Rica report the issue. I had just never heard about it before. Because I worked in the rainforests of West Africa, I mostly read papers about African forests and, apparently, no studies on mist were available for the mountains in Africa.

Like most scientists, I like numbers. I wondered how could I quantify how much less mist was there now compared to before? Meteorological stations tend to measure rainfall, temperature and maybe wind. As far as I know, nobody measures mist. And while images from satellites may show clouds, it's hard to tell if these clouds are touching the ground and are therefore mist, especially if there are other layers of clouds higher up in the atmosphere.

Unexpectedly, the answer to all my questions came to me a few days later.

As I was finishing my interviews in another Samburu village, the son of the village chief told me, 'Since that small plane crashed because of very thick fog, I think the man working at the airstrip records fog presence, at least the thick fog which does not allow planes to land.'

Really? Someone writes down if it is too foggy for a plane to land? I did not even know there was a small airstrip around here!

The next morning I went to the small airstrip looking for the man in charge. Empty. I went again the next day at a different time. And the next day. Eventually I found him. And, indeed, he had been recording whether there was 'thick fog' or not a few times a day for many years, in a notebook. He showed me the notebook for that year. I asked if the old notebooks might have been kept somewhere.

'In theory yes, at the headquarters in Nairobi,' he said.

So, the answer was 500 kilometres south.

Two months later, I had finished measuring trees in the forest on Mt Marsabit and had completed more interviews. Altogether, I had spoken with the elders of 10 villages around Mt Marsabit and in all the interviews the elders had told me about the importance of mist.

It was time to look for the records of the mist and at the Kenyan Meteorological Department an extremely kind lady found a stack of dusty notebooks spanning 30 years for me. After spending nearly two weeks typing the data into Excel, I was able to produce a figure and run some statistical analysis on my laptop in Nairobi. It was a very exciting moment. Was mist disappearance as severe as the old men had said?

I expected to see some change, but I was shocked when I saw the extent of it. There had been a 70 percent decrease in hours of 'thick mist' per year in the last three decades.

Wow.

That is why the elders were talking so much about it! Imagine how much harder it was for the beans to germinate and how much drier the grass was at the start of the dry season! Maybe the lost mist explained why most streams in the mountain had changed from permanent to seasonal. Most likely it could also explain why there was limited natural regeneration in the forest. Without the mist, the forest was drying out and would eventually disappear.

I realised the forest on Mt Marsabit could be a 'cloud forest'. Although it did not look like the ones I had seen pictures of before, it actually did have some similar characteristics. Stereotypical cloud forests—which I had only seen pictures of before—are dark forests where trees are short and twisted. Moss is everywhere, on the ground, on the tree trunks and on the branches. They are like the forests I always imagined in fairy tales, where a gnome could easily cross your path.

But after reading the scientific articles in the cybercafé, I had learned that there are many types of cloud forests, some which are a bit drier and have less moss—possibly like the one on Mt Marsabit.

These wondrous, strange-looking forests create, and maintain, their own microclimate. The leaves and branches of trees, together with the moss, intercept the tiny water droplets from the mist. Many of these tiny water droplets drip into the soil and gather to become small streams or are used by plants to grow. But some of these tiny water droplets intercepted in the canopy of the trees will evaporate and form new clouds which, when touching the ground again, will again form mist. These mist water droplets might be intercepted by leaves and branches, starting the mist-water cycle again in a fascinating natural process.

All cloud forests are dependent on clouds or mist for survival. Some scientists predict that cloud forests will shift to higher elevations with global warming—if there is land available. This would not be an option for Mt Marsabit though, as the forest is already only found near the mountain top; the forest begins at 1,500m, and the mountain is only 1,707m. So, with increased global warming the forest—which now covers about 11,000 hectares—will one day just disappear, with terrible consequences for plants, animals and the Samburu people who depend on it.

The urgency of the situation was clear. We needed to move fast to get a better understanding of the ecological functioning of these forests before they disappeared. After I left Kenya and returned to the UK, where I was based at the time, I immediately began writing scientific project proposals that would allow me to come back to study Mt Marsabit's cloud forest. It took two years before one was finally funded; it felt like an eternity.

My project aim was to measure the mist in three cloud forests, Mt Marsabit, Mt Nyiro (2,752m), and Mt Kulal (2,285m), which were all populated by Samburu pastoralists. I had learned from other researchers how to sample fog using simple wire harps (square frames with fishing lines which intercept the tiny water droplets from mist) connected to a machine that logs how many times a five millilitre spoon has tipped after filling with water from the mist.

I constructed and installed several of these contraptions on each mountain. This is easier said than done. I needed to select a natural gap in the forest, large enough that the airflow wouldn't be disturbed by trees, and the sample of mist wouldn't be skewed by water dripping from the branches of trees. The area also needed to be relatively accessible as I needed to return every few months to download the data and change the batteries. I would be travelling to Kenya every three months for the next two years.

On Mt Marsabit, where the slopes are mild, it was relatively easy to find a suitable spot for my equipment. Mt Nyiro and Mt Kulal both have steeper slopes and are more remote, which made this a real challenge. To access my equipment on Mt Nyiro each time took nearly a full week trip; a day of driving from Marsabit town, crossing a dusty non-tarmac road through the Chalbi Desert in four-wheel drive, a day of hiking up the steep mountain slopes to find the first gauge and then another two days to walk along the ridges of the mountain to check the other gauges.

I contacted a remote sensing specialist—a colleague of a colleague—who was an expert in the analysis of satellite images and who was interested in the dry cloud forests of northern Kenya. Combining my field measurements of mist with satellite images and modelling techniques, we quantified how important mist was for forest dynamics. In short, the greener the forest, the more rainfall it needs. We used satellite images to determine how green the vegetation in the forest was and then gathered rainfall estimates from another type of satellite imaging. We then compared the two and saw there was 'missing precipitation' which should be the part provided by mist. And then we were able to use my field measurements of the mist to determine if the amount matched what was predicted from our estimates. By then I had 18 months of field data on mist to work with.

The data matched. The forest was reliant on mist. I was actually surprised to see that it matched so well, because the complexities of forest-climate relationships—like anything in nature—are difficult to disentangle. This was not a controlled lab experiment where you can be sure of air temperature, water used to irrigate and soil characteristics; this was an analysis of field measurements collected in the real world, where many factors interplay. It could have been a particularly dry—or wet—year and mist measurements may not have been representative of inter-annual mean values. But actually, they matched perfectly.

This gave us the information we needed to predict future changes in forest extent related to predicted changes in climate. When I looked at the figures, the news was devastating. The forest would disappear with just two degrees of warming—an increase we are almost guaranteed to have. And some climate projections predict four degrees of warming, not two.

It was all condemned to disappear. The beautiful forests, the huge trees and the tiny orchids, the wild animals that depended on them, including a species of chameleon only found on one mountain in the whole world, and the Samburu people and their way of living. With no forest left, there would be no permanent water, no place to graze their cows. How could I explain to all the lovely people I had met that in the near future their children would be living in a slum in Nairobi? I felt overwhelmed by what we'd learned.

What could we do about it? We could water the seeds and seedlings of the tree species in the forest to help them germinate and grow. Yes, it would be expensive and complicated to organise, but at least in Mt Marsabit, where slopes were less steep and there were more people living nearby, it could be done. At least if there were still lots of trees and the canopy was closed, only a little sunlight would reach the forest floor and the forest would stay a bit cooler.

But this was unlikely to be enough, as over time even large trees which have deep roots would suffer from the lack of mist. The quantity of water coming into the forest would be so much lower without the mist that the ecology of the whole forest would change. Trees and other plants which require more water would die and, eventually, only tree species which can survive in hotter and drier conditions (often deciduous trees) would be left. Moss would also dry out and disappear and some orchids and other plants, and even certain amphibians, insects and birds, would no longer be able to survive there. As the forest became drier, it would also become more prone to fire. Fire destroys trees whose bark is not resistant to it, so over time recurrent fires would change both the floral and faunal composition of the forest. It is a process which cannot be reversed. In ecology we call this 'going beyond a tipping point'—the ecosystem (in this case the cloud forest) would not be able to return to its original state (a mist-water dependent forest), even if the mist was to come back.

And this process is not restricted to Mt Marsabit, Mt Kulal or Mt Nyiro. Most mountains in Kenya contain cloud forests, which collect water droplets from mist, contributing to creating and maintaining large rivers in the country. These rivers are important for agricultural irrigation, for electricity production, for inland fisheries, for tourism purposes and for providing enough drinking water for domestic use in major cities. And there are a lot of people living in Kenya—more than 50 million. Cloud forests are not just important for Kenya's rivers, water resources and economy, they are also important to Ethiopia, Tanzania, Uganda, Rwanda, Burundi, Mozambique... In fact, they are important in many other countries across the world as well.

After this research, I wrote two scientific publications on the cloud forests and presented the findings at international conferences. I gave a few talks at relevant national organisations in Kenya, such as the Ministry of Forestry, Kenya Water Tower Agency and Kenya Meteorological Department, plus at international organisations such as the Center for International Forestry Research and Mountain Research Initiative. I also wrote several more research proposals to continue the research on cloud forests in other mountains in Africa, using a science-with-society approach, which means combining indigenous knowledge from local peoples and scientific measurements, as I did in Mt Marsabit. And, importantly, I brought my findings back to the Samburu communities. At the end of the day, I had only discovered the drying out of cloud forests because of their comments.

Many parts of the world are data deficient. In climate change research, this means that there are few (if any) functioning meteorological stations and little historical data that can be used to calibrate models and remote-sensing products. Some of these include conflict areas or are just very remote and have historically been little visited by scientists. In such places, indigenous knowledge (the traditional knowledge local communities hold) can help scientists understand the climatic changes observed and the impacts these have had. Yet including indigenous knowledge in climate change research is not mainstream. My study on mist and cloud forests contributes to the slowly growing body of scientific evidence that showcases the importance of including indigenous knowledge.

When I returned to Kituruni in Mt Marsabit three years after that first interview, I sat under the same tree by the dusty road where I had first heard about the mist and spoke to the elders once more.

'The forest is drying out, yes. God is angry with us because we changed everything,' said the old man. And the other old men nodded.

Once more, a heated discussion followed, which again I could only follow here and there thanks to Mohamed my field assistant. Luckily, I had brought my water bottle this time too. The once nomadic Samburu pastoralists of northern Kenya had changed considerably in the past few decades, but looking at this on a larger scale, we humans had changed so many things on our planet. Population growth, industrialisation, deforestation, land use change, non-organic pollutants...

Once more, the old man made me wonder. Was Mother Nature angry at us humans for everything we had been doing? Can we really reverse what we are doing? I think we can—but do we have the will?

'WE HAVE LARGER PROBLEMS THAN CLIMATE CHANGE'

BY
AIDA CUNI-SANCHEZ,
WITH GUIDANCE AND INPUT
FROM GHISLAIN K.R. BADERHA

Ghislain Baderha *completed his Master's degree at the Department of Biology, Université Officielle de Bukavu in the Democratic Republic of the Congo. His interests focus on sustainable use of natural resources and local communities' adaptation to climate change. He works as a researcher and independent consultant.*

I am cold, hungry and extremely tired. 'Why am I here?' I ask myself.

I am inside a tiny hut made of mud and straw. Sitting on a small piece of wood, by the side of some smelly goats—and their poo, of course—Ghislain and Rodrigue, my Congolese master's students, are sitting next to me. It seems that I will be allowed to spend the night on the little bamboo bench, my luxury as the only female in the group. My students will have to share the floor with the goats. I feel so sorry for them. I can see that it's going to be a very long night.

Maki, our guide and armed guard, is trying to convince our host to cook some food for us. But the old herdsman refuses. It might be that he does not have firewood or water left and it is too late to go and fetch some. Or he might be afraid to light a fire after darkness. Once darkness falls, one should not wander around in this mountain full of rebels. The latest estimates say there are 14 different rebel groups fighting for the control of the different artisanal mines, so wandering around at night is a bad idea. Instead, we eat some peanuts and biscuits for dinner. Luckily, the host gives the others some fresh milk, but unfortunately for me, I am allergic to milk.

We have been hiking all day up this mountain, following the small footpaths up and down through the forest, the grasslands, the swampy areas and the rocky escarpments. Carrying too much and too little at the same time. Trying to be fast, careful and unnoticeable, the latter challenging given the white colour of my skin. We are going to the famous Lake Lungwe, the sacred lake near the top of the mountain. We want to talk to the Nyindu people living next to it. This is easier said than done, of course.

From the last village with car access, it takes two days to hike there—if you are young and fit. Otherwise, maybe three days. And it is not so safe. You must travel only during daylight hours. And you must always keep asking local people about the security situation. If they say it is not a good time to go, you turn around and walk back home. Last time I was here, two years ago, someone killed the village chief of the village where we had planned to spend the night. So, we turned around and went back to the city. I hope the rebels have no issues to settle with the villagers this year and we can safely access the area and meet with the Nyindu people.

I am involved in a small research project once again documenting climate change impacts in African mountains. We are interested in understanding how rural farmers in mountain regions experience and adapt to climate change. In areas with good access to urban markets, farmers might use improved seeds, fertilisers and pesticides to boost crop yields. In areas where farmers have access to reliable weather forecasts, they might adjust their planting dates. Some farmers might invest in irrigation, some in integrating trees into their farms to provide more shade for coffee; some might decide to abandon farming and engage in other industries, such as tourism.

But what are farmers in the mountains of Eastern DR Congo doing? Here there are few NGOs, no assistance from the government, no access to roads, no mobile phones to view weather forecasts, no tourism to provide other employment opportunities. What do they do? To find this out is the crux of Ghislain's master's thesis. And getting answers from 150 farmers is going to be challenging, as first we need to climb to Lake Lungwe to find them.

The night is indeed very long and Ghislain and Rodrigue sleep very little. In the morning I see the first signs of bed bugs in my skin; if they have infested my sleeping bag, the coming nights will also be long and painful. There's nothing I can do about that now though.

Our host makes us some warm tea and I am so happy! We then start the hike again. Up and down, through another patch of forest, another grassland, a patch of bamboo forest—we say hello to another surprised herder. I am sure I am the first white lady he has seen in his life. Around 3pm we finally see the famous lake. It is indeed magical, its waters reflecting the sun amongst the mist. Three shy ducks fly away because of our presence. No more sound, wind or movement. Beautiful as a postcard.

After meeting the village chief, the elders and a few more people whose names I no longer remember, nor their roles in the community, we are given a house to spend the night. Apparently, it is not safe to spend the night in tents here as the strange colourful structures could attract unwanted attention. Okay, another bamboo bench for me tonight. Our guide's sister agrees to cook dinner for us, and the long ceremony of fetching firewood and water begins. I just sit in a rock by the entrance of the house, looking at the stunning sunset over the lake waters. We finally made it. Tomorrow we finally start the survey—I hope.

'The rains have changed indeed. They no longer come when you expect them. And when they come, they are intense, but they go fast,' says one respondent.

'They are also less consistent, sometimes, the rains stop for a while and the maize stop growing,' adds another one.

'The main problem is that now there are rains during the dry season, so the maize cannot ripen correctly,' says another one.

'Because of the changes in the rains, there are now more pests in maize, such as that worm [the fall armyworm],' another states. 'And the cows are also sick more often.'

'So what have you done to adapt to these changes?' asks Ghislain.

But most respondents say the same: not much. There are no NGOs giving improved seeds, such as fast-maturing varieties of maize, or varieties more resistant to the pests here. Buying pesticides is expensive and carrying bags of them uphill for two days is extremely challenging, so they are not used. It's the same for fertilisers. Investing in things which take time to show results, such as agroforestry, irrigation or terracing soils, is not common because of the insecurity.

'How can you invest in your land if you do not know if you will be living in the same village next year? We have bigger problems than the changing rainfall,' says one old man.

I found myself nodding. I feel for these people; their lives are very difficult in these remote mountains. I have no answers.

'This is our land, my ancestors were farming here, my grandfather was farming here, my father and myself. But I am not sure my children will be able to do so,' he adds. 'And this is not because of climate change.'

Indeed, you cannot invest in your farm or even start a new venture, if you do not know if you will be on the same land next year. Or if your shop or house will be burned and you and your family forcibly displaced. High uncertainty makes planning difficult, even nonsensical. Even increased intensification of agriculture only makes sense if you can sell agricultural surplus to a market. And here these are out of reach. Some respondents said that they send their children to school, so that they can get a job in the city in the future and send money back to those living in the village. Indeed, sending children to school is a climate change adaptation strategy, but one focused on mid-term adaptation. It won't help in the short-term.

Some respondents also say that they are turning to mining, to diversify livelihoods, but mining is a risky choice. Miners work for days, weeks and sometimes months before they find high-quality material. And even when they do, they get paid little money for it, as most money is made by middlemen or traders. Miners often spend most of the money they make in mining on site, buying nice food, alcohol or prostitutes. Little money reaches their homes. From the conversations we have, diversifying livelihoods into mining does not seem to be an effective or positive adaptation strategy.

That evening the headmaster of the only primary school for kilometres around comes to visit us. He complains about the lack of furniture, books and students—several families have left the area already. The discussion moves into the potential loss of local culture, traditions, the Nyindu language... As people move out and into the villages of other ethnic groups, over time they will 'blend with them'. Is their distinct culture condemned to disappear?

'But what could be done to help you adapt to climate change impacts?' asks Ghislain, and the discussion continues in Swahili, a language I cannot follow very well.

After a long conversation, Ghislain summarises the answer for me in two words: not much.

Really? Maybe they are right. Not much. How can you build anything in this insecurity?

The next two days we make lots of progress, in terms of people interviewed. In terms of responses, not much is new. In short, local farmers have observed increased temperatures and changes in rainfall distribution and amount, but they have changed their farming practices very little. And they have little idea of what else can be done.

That evening there is a full moon, and the lake is glowing. 'Beautiful,' I think.

As I am brushing my teeth by the door of the hut the village chief's son comes to visit us. He comes alone, strangely. He tells us that things have changed and that his father recommends that we leave the village by sunrise. I ask if Ghislain can stay to finish the remaining surveys without me; as I am the only white person, I am more conspicuous here. We have only completed 132 surveys; we'd like to reach 150. But he says leaving Ghislain behind might be too risky as he does not speak the Nyindu language.

So, we pack and say goodbye to our hosts before going to bed. That night, two armed guards as well as our local guide stay by the entrance to our hut. One can feel things getting tenser. I sleep little, like the others in the team. You should never take a village chief's advice lightly in these mountains.

On the way back down, I find myself again sitting in the same tiny hut with the goats as on the way up. I'm tired, cold and hungry, but I am slightly happier than last time, as I have seen the lake and helped interview more than 100 Nyindu farmers. But the mood is not that of a celebration. We are still on the 'unsafe' mountain and the village chief told us to go back to town as fast as possible. But we simply cannot walk as fast as the locals here. It also rains on the way down, the path muddy and slippery. We try to take it easy. I don't want a twisted ankle here. Or a broken arm.

This time my colleagues convince the old herdsman to let them cook some food for dinner—maize porridge with milk. The sugar is finished. I can only eat the porridge when it's made with water anyway. Not sleeping is one thing, but not sleeping and being hungry is even worse. It is even colder than last time, and I spend the whole night dreaming of a hot shower. But, I tell myself, tomorrow we will be back in town and maybe I can ask the lady at the guesthouse to warm some water so I can shower. Maybe we can even eat some rice and vegetables... I am tired of the maize porridge.

I think about the herdsman who lives here with his goats. He will not go back to town tomorrow. He will be here, eating maize porridge, drinking milk, fetching firewood and water and going back to the hut before it gets dark, for safety. He wants to be here for as long as he can—if no rebels come and burn his hut. The people we interviewed in the mountains were right. Some people have more pressing issues than how to deal with climate change impacts.

RAINDROPS KEEP FALLING ON MY HEAD

BY

KEVIN E TRENBERTH

Dr Kevin Trenberth *is a climate scientist. He is a Distinguished Scholar at the National Center of Atmospheric Research in Boulder, Colorado, and an Honorary Academic in the Department of Physics at the University of Auckland in New Zealand. From New Zealand he obtained his doctorate from Massachusetts Institute of Technology. He has been prominent in most of the Intergovernmental Panel on Climate Change (IPCC) scientific assessments of climate change and has also extensively served the World Climate Research Programme (WCRP) in numerous ways and many US national committees. He is a fellow of the American Meteorological Society, the American Association for Advancement of Science, the American Geophysical Union and is an honorary fellow of the Royal Society of New Zealand. He has published over 590 articles and books and is one of the most cited scientists in his field. More than anyone else he has promoted how extremes of climate arise from climate change. His most recent book is* The Changing Flow of Energy Through the Climate System, *2022, Cambridge University Press.*

Water, water, everywhere
The rain is pelting down, and visibility is very low even in the middle of the afternoon. Water is all around on the ground and flooding is obviously widespread. At our house, water is flowing into the basement which is flooding. An examination that got me thoroughly drenched reveals a blocked downspout as well as several downspouts emptying their huge deluge from the roof onto the lawn where it is circulated back to a window-well. The latter is one metre deep and full of water; the adjacent window is leaking vigorously into the basement of our house. Climbing a ladder with a broken foot in pouring rain is not recommended but it is what it takes for me to free up the blocked downspout, thereby redistributing the load. Then deployment of some innovative plastic and pipes redirected the downspout flows away from our house (no doubt to the guy next door). But it eased, and even ceased, the flow of water into our basement.

This was the day, Thursday, September 12, 2013, when Boulder, Colorado, smashed all precipitation records with over nine inches of rainfall. It was my fourth time caught up in an extreme weather event—and one I will cover in more detail below.

Climate scientists can tell weather (or not)!
I began my scientific career long ago (1966) in the New Zealand Meteorological Service. For a period, I became a junior weather forecaster, doing shift work. At the same time, I was playing senior level rugby (sometimes against All Blacks, members of the NZ national team). Watching weather go by, I perceived certain

weather patterns that affected rugby practice and games. I was fortunate to win a New Zealand Research Fellowship to study for my doctorate at MIT, but on my return to New Zealand, I explored from whence those patterns arose. Analyses of the rather different weather from one year to the next led to discovery of weather regimes: a quasi-biennial oscillation, plus patterns that turned out to be related teleconnections (global-scale wave patterns) associated with the El Niño Southern Oscillation (ENSO).

At that time no-one was working on that topic, but I published a paper in 1976 on my findings that became sort of a classic. I moved with my family from New Zealand to the United States in 1977. Interest in El Niño blossomed soon thereafter, and I was invited to be on a new NOAA committee to explore the topic along with extensive field work (using ships and buoys). In due course, I and others proposed and became involved in the international TOGA (Tropical Oceans Global Atmosphere) Program that was set up in 1985 to explore ENSO as the first project under the new World Climate Research Programme. Climate change was not a public concern at that time, but an ad hoc National Academy of Sciences report chaired by Jule G. Charney was published in 1979 saying it should be. It was not until the 1980s that concern became great enough to establish the Intergovernmental Panel on Climate Change (IPCC) whose first report came out in 1990.

As a junior weather forecaster in New Zealand, I had gained experience in answering queries from the public about all sorts of things related to the weather. It led to my approach in how to explain climate variations. The framing I have always intuitively used is to recognise the huge chaotic variability of weather systems, but also to recognise the maybe small but systematic influences of effects external to the atmosphere, especially the oceans. Fundamentally, if all that is happening is atmospheric variability, then there should be no reason for systematic patterns to occur. Accordingly, when persistent patterns exist, it is a sign of systematic forcing of the atmosphere, the most notable example being El Niño.

Global climate change has only become strongly evident since the late-1970s. As the climate has changed, so too has the environment for all weather systems. Most important are the changes in temperatures in the ocean and atmosphere, and associated changes in water vapour in the atmosphere. The oceans are warmer; accordingly, sea levels are higher and the air over the oceans is warmer. Because the water-holding capacity of the atmosphere depends strongly on temperature—it increases seven percent per degree Celsius—there is also a direct relationship with humidity and there is now more moisture over the oceans by five to 15 percent compared with pre-1970 values. Accordingly, as cyclonic weather systems reach out and bring the low-level moist air into the system, precipitation intensity increases. It rains harder! That in turn releases latent energy into the weather system, potentially making it more intense, bigger and longer lasting.

Of course, none of us experience global climate directly. It is one thing to write about what we think is happening to weather in association with weather regimes, teleconnections and climate change, but it is through each individual event that we experience it.

And, it turns out, I have experienced more than my share of disasters: the *Wahine disaster* in Wellington, New Zealand, 10 April

1968; *Lower Hutt flooding*, New Zealand, 20 December 1976; *Superstorm Sandy*, New York, 29-31 October 2012; and *Record flooding in Boulder*, Colorado, 9-16 September 2013. The first two were before climate change was a factor, but the last two were clearly made worse by climate change.

The *Wahine disaster* was caused by a rapidly developing hybrid storm (ex-hurricane with extratropical influences) with local winds up to 168 mph in Wellington. I was caught out in the storm and my small Austin car had the paint sand-blasted off in places. I reached the Met. Office about 10am just as peak wind gusts of 123 mph were recorded there, and I had tremendous difficulty getting from where my car was parked into the building. Four times I rounded a corner of the building to make a beeline for the main entrance, but I was knocked back.

I was fortunate to be able to observe a lot of subsequent developments from the Met. Office, about 150 metres above the main city, in relative safety. Debris and bits of roofs were flying all over the place. The inter-island ferry, *Wahine* (a roll-on, roll-off ship 150 metres in length) carried many cars and had 610 passengers and 123 crew on board. Fifty-three people lost their lives when the ship foundered on Barrett Reef at the entrance to Wellington Harbour and capsized. The wrecking of the *Wahine* is by far the best-known maritime disaster in New Zealand's history. On the same day, some 98 houses in Wellington lost their roofs. Insurance industry payouts exceeded $200 million (2008 NZD) on 3,657 claims. While climate change was not factor, the hybrid nature of the storm was a preview of *Superstorm Sandy*.

In 1976, my family and I lived in Normandale, a suburb on the western hills overlooking Lower Hutt, New Zealand, about 20 kilometres northeast of Wellington. *The Lower Hutt flooding* on 20 December was a rare rainfall event that dropped 280mm of rain in our neighbourhood. The weather situation produced very high levels of water vapour in the atmosphere, a forerunner to a climate change signal in more recent floods. Many slips pushed houses off foundations, destroyed some houses and cars and caused power failures. Ground slope failure and mud flows were common, and it was estimated that there were some 925 slips in the area. Over 100 families were evacuated.

It took me many hours to get home to my wife and two-year-old daughter that day from the Met. Office, and it was fortunate that there was another indirect route to our house near the crest of the hills. While our house was sound, it was located near the top of a fairly steep hill, and a huge chasm opened up from erosion on the next property to the south. Our house was cut off from Lower Hutt for weeks and damage was severe all around us, right at Christmas time. The cost was estimated as $205.2 million (2009 NZD). As of 1995, it was New Zealand's most expensive flood.

Superstorm Sandy caused tremendous damage when it made landfall on the New Jersey coast and New York area on Tuesday, 30 October 2012. It began as a hurricane but became a huge hybrid storm as it moved erratically north before making landfall. The worst problems on the Jersey Shore were caused by the strong winds and the associated storm surge, leading to extensive flooding. Farther inland, heavy precipitation was also a major problem. Widespread damage from flooding streets, tunnels and subway lines plus cuts of power in and around New York City led to damages exceeding $65 billion (2013 USD) and more than 110 lives were lost in the New York and New Jersey area.

Superstorm Sandy was the first disaster I had a personal connection to where climate change was a factor. Climate change made the storm much more powerful than it otherwise would have been-and made it clear that our warming climate was changing the scale of what humankind can expect from 'natural' disasters.

I was involved in two different ways with *Sandy*, one professional and the other very personal. I was in Boulder at the time. From there I wrote several popular articles that were published in many places, as well as a scientific article on the role of climate change in *Superstorm Sandy*, and I spent several days in dealing with the media, including appearing on NBC Nightly News and local TV. Meanwhile my wife Gail was visiting my daughter and other family in Hoboken, New Jersey: ground zero.

I had alerted them of the prospects for a severe storm a week ahead based on the ECMWF forecast, which was excellent. They prepared well, given they had a garden level apartment with the first floor below the road, but they were evacuated. The area lost power for a week.

Fortunately, their place was okay as sandbags kept the water out. A week later my daughter and family returned home, but Hoboken was still in terrible shape with nothing in stores, some roads closed and no train to New York (through tunnels underneath the Hudson River). Commuting to New York City, where both my daughter and her husband worked, was not possible as the New York financial district not only had no power, their backup was gone, and they also had no heat.

Record flooding in Boulder: 9–16 September 2013

Yet another major flooding disaster—and one also shaped by climate change—was the one my family I experienced in September 2013, at our home in Boulder. Boulder lies about

40 kilometres northwest of Denver at the base of the foothills of the Rockies, at 5,400 feet (1,646 metres) above sea level. Consequently, it can be subjected to upslope rains: the wind direction matters a lot.

In this case, I was on travel in Europe as the storm began and I strived, in the middle of the event, to return home where my wife, Gail, was struggling to deal with some basement flooding in our house. On Thursday September 12, there was over nine inches (23 centimetres) of rainfall in Boulder and the eight-day total was over 17 inches (43 centimetres), both roughly double previous high values.

Starting on September 11, 2013, a slow-moving cold front stalled over Colorado, clashing with warm humid monsoonal air from the south. The situation intensified on September 11 and 12. Boulder County was worst hit, although the whole front range was affected. At least eight deaths were reported and more than 11,000 people were evacuated. Many buildings suffered severe damage and over 19,000 homes were damaged to some extent.

Throughout the storm, it was known that rain was falling, but amounts were grossly underestimated because the radar reflectivity gave wrong answers, owing to the tropical nature of much of the heavy rains. Accordingly, those monitoring the rainfalls greatly underestimated the flood risk. In Denver the three highest atmospheric total column water vapour amounts ever recorded for September (since 1956) occurred on 12-13 September 2013 (as high as 34 millimetres). This may not seem huge, but recall Denver is a mile (more than 1,600 metres) above sea level.

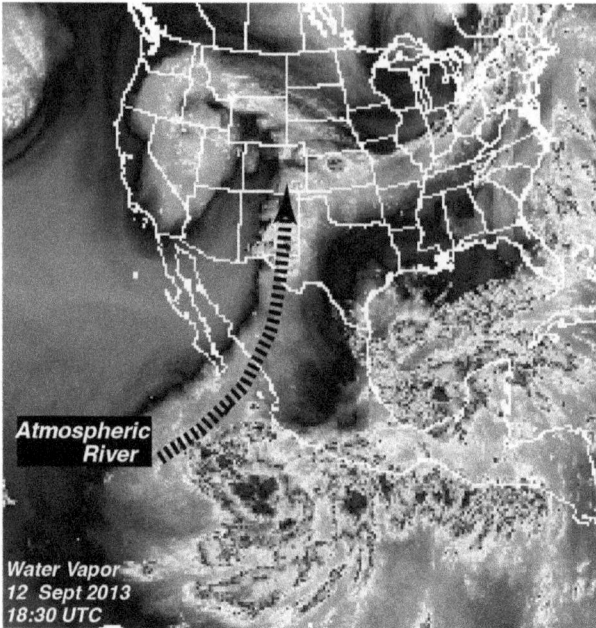

Fig. 1. Water vapour channel imagery for 18:30 GMT on 12 September 2013 showing the extensive water vapour and associated activity both west of Mexico and in the Caribbean Sea and the river of moisture from south of Baja, Mexico to eastern Colorado. Adapted from NOAA.

It so happens that the sea surface temperatures (SSTs) off the West Coast of Mexico, south of Baja, west of Guadalajara, were over 30 degrees Celsius—more than one degree Celsius above normal—in August 2013. This made it the hottest spot in the ocean in the western hemisphere. An incredible 75 millimetres of total column water vapour was recorded in the atmosphere in that region by NASA satellites.

The high SSTs led to the large-scale convergence of moisture flowing that was siphoned north by a very unusual synoptic situation. This led in turn to a river of atmospheric moisture flowing into Colorado. After that atmospheric river shut off, twin tropical storms formed both sides of Mexico: Manuel (to the west) and Ingrid (to the east). This created a double whammy for Mexico, leading to hundreds of deaths, tens of thousands evacuated, tens of thousands of homes damaged and billions of dollars of damage.

SSTs have been as high in the past in this region west of Mexico, but in previous cases they were part of a much larger-scale pattern associated with El Niño events such as in 1997-98, 2004-5, 2009-10. What seemed unique in 2013 was that this was the warmest spot in the Western hemisphere and hence this was the preferred location for low pressure to form and low-level wind convergence, which brought large amounts of moisture into the region. Hence there was clearly an internal variability component to the patterns of SST. At the same time, however, the overall increase of global SST associated with global warming occurring on multi-decadal time scales was also a factor.

Meanwhile, I was at a conference in Graz, Austria, and on the last day (Wednesday) the conference finished a bit early and some of us were taken on a journey to a castle that was somewhat nearby—only 50 kilometres away. We were nearly there and stopped to take pictures of the castle in the distance when I had an accident. I failed to see a curb, fell heavily, and broke a bone in my foot. I did not realise it was broken at the time and so we went to the castle and walked quite a long way (hobbled for me).

I had no time for treatment as I flew back to Denver partly overnight via Frankfurt. A colleague, Bill Kuo, was on the same flight and we arrived back on Thursday in the middle of the major storm during the heaviest rainfalls. We were unable to get a bus to Boulder but shared a cab ($100; double normal costs) and we barely made it into the city before route US36, the main highway between Boulder and Denver, closed. The cabbie was probably stuck in Boulder unless he had the wherewithal to make it out via South Boulder Road. Bill dropped me off and I arrived home to flooding and a deluge.

The back yard was flooded and so was the basement, with about two to three inches (five to seven centimetres) of water, although the sump pump worked. Hence, I had to climb the ladder (all with a broken foot) to clear the downspout (in pouring rain) and reroute and extend the downspouts at ground level to redirect the water from the roof and prevent it from circulating into the window wells. I got soaked, but it worked. Gail and I used a water vac and sucked up the water in the basement to get that under control. The basement was unfinished and so the damage was minimal.

The next day, Friday, the rain eased but devastation was everywhere. Masses of water lay where it shouldn't, many roads were washed out and there was much debris; at least eight deaths were reported and more than 11,000 were evacuated. For some it took years to recover.

There were no Urgent Care facilities open. Boulder Medical Center was entirely closed and Boulder was effectively shut down. I was eventually able to get seen in the emergency ward of the new Boulder Community Hospital at Arapahoe and Foothills although they also had some flooding. They confirmed that the foot was broken and put me in a 'boot'. I was subsequently x-rayed every three weeks and I wore the damn thing for 12 weeks. Then my foot would not work; it was swollen and needed physical therapy. I took trips to London, Beijing, Wichita KS and

New Haven CT (Yale) wearing the boot. My experience finding medical assistance for my broken foot made it starkly clear to me how difficult it could be for other natural disaster survivors to access the treatment they need.

As well as my own published study on this event, there have been several others, with disparate conclusions regarding the role of climate change. Some studies note the unusual weather situation that led to the unprecedented atmospheric river into Boulder and concluded the cause of the event was natural variability. Indeed, the exact setup, no doubt, was dominated by chance. But that statement says nothing about the role of the record high SSTs and the fact that climate change was responsible for at least a significant fraction of that. It is important to ask the right questions. None of the model-based studies were able to replicate the event and part of this is because of inadequate model resolution to be able to represent the structure of the Rockies, in particular.

The most definitive study was not published until 2017. It used an approach that specified the weather pattern and ran a high-resolution model multiple times with and without changes in SSTs from climate change, estimated elsewhere. The result was that anthropogenic drivers increased the magnitude of heavy northeast Colorado rainfall for the wet week in September 2013 by 30 percent. Locally these effects were further amplified on the ground as runoff waters drained into channels and rivers.

Do something!
Disasters of course occur without climate change being a factor, but climate change influences typically make them worse— sometimes much worse. In most respects, my family and I came

through the above disasters relatively well compared with many others. For us, it was mostly bad luck: being in the wrong place at the wrong time. But clearly some places are more vulnerable than others, especially as climate change rears its head. In fact, climate change concerns are one reason my family and I now live in New Zealand. It is also prudent to consider the flow of water around and through any place one chooses to live in, as I have done.

It is also important for communities to take measures to build resilience to any possible event. This is often done in wealthy countries, and the available infrastructure can help enormously when a surprise event does take place. But it is not always the case, and I have been surprised at how poor the infrastructure is in many places, largely because of the failure of communities to tax themselves to build adequate drainage systems. This happened in Houston with Hurricane Harvey in 2017 and Florida with Irma (also 2017) and Ian (2022).

The ability to prepare for disaster is also unfortunately much less the case in developing countries. And too many people, often with few other options, are building on flood plains, such as in Pakistan, where they were badly flooded in 2022. Many developing countries and small island states have not contributed much to the problem of climate change but are suffering disproportionately from the consequences.

Given that climate change contributed to the Pakistan disaster by increasing rains by up to 50 percent, and that over 1,600 people died and more than two million homes were destroyed, the question of reparations arises. At COP27 in Egypt in November 2022, 'loss and damage' was one major concern that received overall support, although it is unclear how well it will be funded and how it would work.

Human-induced effects through increases in heat-trapping gases in the atmosphere continue, and warmer oceans and higher sea levels are guaranteed. As we have seen in 2022, whether from drought, heat waves and wildfires, or floods and super storms, there is a cost to not taking action to slow climate change, and we all are experiencing this now.

According to a quote from the late 19th century, often attributed to Mark Twain, 'Everybody talks about the weather, but nobody does anything about it.' Now humans are changing the weather, and still nobody does anything about it!

ICE IN THE CLIMATE SYSTEM:
A CAREER INVESTIGATING LINKAGES AND CHANGES

—— BY ——

IAN ALLISON

Dr Ian Allison AO AAM FAA is a glaciologist who has studied ice and climate for over 50 years. Most of his scientific fieldwork has been in Antarctica where he participated in or led 25 research expeditions while employed by the Australian Antarctic Division. He has also worked on glaciers in the sub-Antarctic and on equatorial glaciers on the high mountains of New Guinea. Ian has published over 130 peer-reviewed papers on topics including ice shelf-ocean interaction, Antarctic weather and climate, sea ice and the mass budget of the Antarctic ice sheet. Ian has been actively involved for many years in international collaboration in Antarctic and climate science. This includes serving as a Lead Author of three of the Intergovernmental Panel on Climate Change (IPCC) Assessment Reports, and as the co-Chair of the Committee that coordinated the International Polar Year 2007–2008, a two-year research program involving more than 130 internationally collaborative science projects, and 50,000 participants from 60 countries. His work in research and science coordination has been recognised through both national and international awards. Since formal retirement, Ian has retained an abiding interest in research into ice-climate interaction and has continued involvement in ice and climate research through his honorary association with the University of Tasmania.

Introduction

This chapter is largely a personal recollection, based on my own experiences, of how research into the cryosphere (a collective term for all forms of ice on the Earth's surface) has transitioned over the last decades from a subject of exploration and basic curiosity to one of critical importance for understanding Earth's climate system, its future changes and their global impacts. This applies most particularly to the role of the Antarctic and Greenland ice sheets in future sea level change.

How did I end up in a career as fascinating and, back when I began, as unusual as glaciology? After high school years that were academically successful, but not at all oriented to team sports (although I had an early interest in skiing and wilderness walking), I commenced a science degree (physics and applied mathematics) at the University of Melbourne. By my third year, when applied mathematics had largely dropped off the agenda, I took a one-term unit in meteorology. When I completed my basic degree and decided to continue further study, I chose that applied-physics discipline. I was, perhaps, attracted more by the fact that the Meteorology Department then coordinated Australia's glaciological research program in Antarctica (and that there were often skis on the roof of the departmental vehicle), than by the lectures in atmospheric dynamics I had sat through.

In 1966, I completed an Honours degree within the Meteorology Department and then commenced an MSc degree in micrometeorology, studying the transfer of heat from the atmosphere into the ground. This left me as a strong contender for a scientific-expeditioner job with the Antarctic Division to undertake a similar research program investigating the exchange of heat between the atmosphere and sea ice. My application was successful and I began in March 1968.

Antarctic coastal land-fast sea ice

My research program then was about climate processes and not about climate change. I spent 1968 designing and preparing equipment for a study of the growth of land-fast sea ice (ice that forms on the ocean surface but remains fastened to the shore) and its interaction with the ocean and atmosphere, over a full seasonal cycle. I sailed for the Antarctic in December 1968, spending all of 1969 based at Mawson station. To study the full seasonal cycle, I needed to make measurements over open water during late summer and autumn, before the seasonal ice cover formed, as well as over the solid winter ice which grew to about 1.5 metres thick. The open-water measurements were made with recording instruments mounted on a raft, crudely constructed from welded fuel drums and scaffolding tubes, and moored downwind from the shore. I made these more difficult measurements over two summers, in both early 1969 and in early 1970. I returned to Australia in March 1970 and spent the rest of that year analysing and writing-up the results.

Sea ice studies became a main focus of my research over the subsequent years. The 1969 Mawson study was developed into a multiyear program to investigate inter-annual variability. Others were recruited to the team to spend a year at Mawson making the measurements. These became increasingly more detailed and more accurate, including oceanographic measurements to look at what was happening in the deep-water column below the surface. The results from this program quantified how the different processes of heat exchange between the atmosphere and the ocean (solar and infrared radiation, convection and conduction into the ice) varied throughout the year and affected the seasonal growth and decay of the ice cover. They also gave an indication of the inter-annual variability in these processes. Plus, they showed that upwelling of a small amount of heat from the weakly stratified ocean to the underside of the ice was also important and reduced the overall maximum ice thickness by over 30 centimetres.

Sea ice, in both polar regions, plays a major role in modifying the heat exchange between the ocean and the atmosphere since it reflects much of the incoming sunlight and because it acts as a physical barrier to heat, moisture and momentum transfer. With a warming climate, the extent of sea ice will decrease and the ice-free ocean will then absorb more solar radiation. This can lead to positive feedback whereby the climate will warm even more. Sea ice growth also modifies the ocean structure. As the ice forms and grows, it ejects salt into the underlying water column, increasing its density so that the surface water can sink to the bottom of the ocean. This vertical transfer connects the deep ocean to the surface and plays an important role in the global climate system. Over the years our research thus became increasingly focussed on the impacts of climate change on sea ice and the resultant feedbacks.

Pack ice in the Southern Ocean
But land fast-ice like that studied at Mawson typically extends only tens of kilometres from the coast and represents less than five percent of the total Southern Ocean area covered by sea ice at the

seasonal maximum in October. Most of this vast area of nearly 20 million square kilometres is covered by pack ice consisting of broken pieces of ice, called floes, of varying size and thickness which are constantly moved by wind and ocean currents. Small areas of ice-free water, called leads, can open between the floes. The pack ice is constantly changing and deforming. Floes less than about 40 centimetres thick can be easily driven on top of each other (a process called rafting); thick ridges build as floes collide and stack broken pieces of ice into piles above, and below, the sea surface; new leads open when the floes diverge, and thin new ice rapidly forms in these; and snowfall and wind-blown drifting snow continually modify the terrain.

To understand the global impacts of sea ice it was thus necessary to make ship-based measurements from within the pack ice and far from the coast. I was able to commence a program of ship-based observations of pack-ice characteristics in the 1986/87 season, and this became much more feasible after the Australian Antarctic program acquired the ice-breaking research and supply vessel, *RSV Aurora Australis*, in 1990. Over the next few decades, my investigations of pack ice characteristics and processes became part of the overall Antarctic Division marine science program, and usually included collaboration with physical oceanographers. Observations from the ship were often extended by helicopter observations.

In 1995, and again in 1999, I had opportunities to conceive, plan and lead the only two Australian marine science research voyages that have worked deep within the sea ice, right to the Antarctic coast, in the depth of winter. The first of these, from July until September 1995, investigated changes in the ocean properties during a time of rapid ice growth within the pack ice. This was undertaken over the Antarctic continental shelf near 64°S, 140°E. *Aurora Australis* traversed three times around a 110x110-kilometre experimental area while oceanographers measured changes in the water structure with instruments lowered through the full depth of the water column. The pack ice drift and deformation were measured with an array of eight GPS equipped drifting buoys, with the data relayed via satellite. We glaciologists determined changes to the ice characteristics and thickness distribution from observations on ice floes, from helicopters and from the ship. Meteorological observations were made from the ship and from two of the drifting buoys.

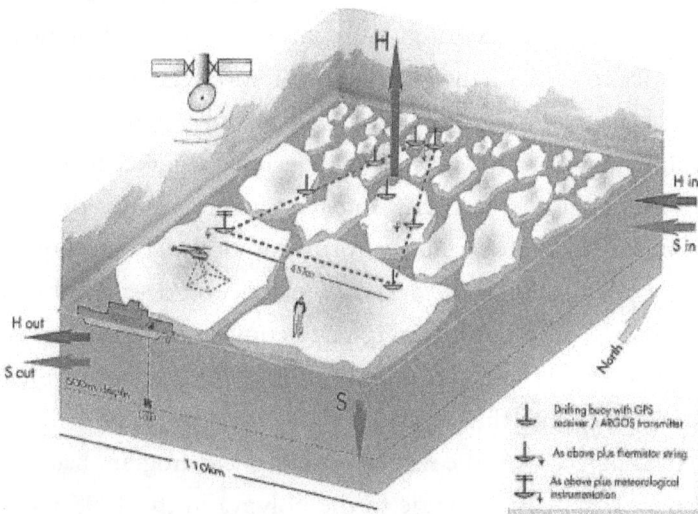

Schematic of the 1995 winter sea ice-ocean interaction study. The arrows labelled H and S represent fluxes of heat and salt respectively.

The second winter voyage was to the Mertz Glacier Polynya (MGP) on the coast of the Antarctic continent at approximately

67°S,145°E. Polynyas (a Russian word) are large areas within the winter sea ice zone that remain ice free, either because upwelling of warm ocean water prevents ice formation, or because strong and persistent off-shore winds remove the ice as soon as it is formed. The Mertz Glacier Polynya is of the latter type. Strong katabatic winds, flowing down the Antarctic coastal slope, sweep the new ice forming in the polynya north and westward. The then 80-kilometre-long floating tongue of the Mertz Glacier blocked ice moving with the coastal winds from entering the region from the east (the tongue has since broken off as a giant iceberg).

Although they are relatively ice free, coastal polynyas are somewhat paradoxically regions of very high ice production. Ice that is formed is immediately swept away by the wind, allowing further ice to form on the exposed ocean surface. Up to ten times as much ice can form each year in a coastal polynya as in the surrounding pack ice zone. The salt rejected during ice formation plays a critical role in increasing wintertime salinity over the continental shelf, leading to the formation of very dense 'bottom water' which sinks to the abyssal depth of the ocean and plays a major role in the overturning circulation of the global ocean. Between 13 July and 7 September 1999, *RSV Aurora Australis* travelled deep into this polynya for a wintertime study.

This, however, was the second time that the program had been attempted. An earlier voyage to the polynya in the 1998 winter ended abruptly with a major fire on the ship. The ship entered the sea ice on 21 July 1998 and, shortly afterwards, in the early hours of 22 July, a buoy was deployed on the ice to track the sea ice drift. Most people retired to their bunk after this, only to be rudely awoken at 2:30am by a clamouring fire alarm and a faint whiff of smoke. A very serious engine room fire had occurred when diesel fuel from a split fuel hose sprayed onto a hot engine turbocharger.

Within 10 minutes of the first alarm, the 54 scientific personnel onboard had assembled at the muster station on the helicopter deck. The temperature was -13°C, but fortunately with only a slight wind. Soon the main power failed, and emergency lighting came on. After all of the 25 crew members had also been accounted for, the engine room fire doors were sealed and the halon extinguishing system was activated, showing as thick pungent clouds from the exhaust stack. The crew prepared the lifeboats and swung them to the embarkation level as a precaution. A Mayday message had been radioed shortly after the first alarm. The scientific personnel remained on the deck for about an hour, before moving into the helicopter hanger for another hour and a half, and then to one of the lounges.

I was fortunate that I had a clear job to distract me from worrying about any personal consequences. My deputies, Tony and Vicky, and I were responsible for ensuring that all the scientific personnel remained assembled in a safe place while the well-trained crew handled the emergency. We tried to keep spirits up and allay fears as best we could. From my briefings with the captain, we kept everyone informed of the developing situation and plans. Crewmen wearing breathing apparatus periodically checked the engine room, but it was not until 2pm, more than 12 hours after the initial alarm, that the engine room was opened for natural ventilation and most people returned to their cabins.

Weather conditions were near-blizzard the next morning as the incapacitated ship drifted WNW within the pack ice at about one knot. Over the next few days, the ship's engineers slowly restored water, sewage, an emergency generator and then a functioning galley. But it was not until 5pm on 25 July that one of the two engines was sufficiently repaired for the damaged vessel to start slowly heading home. Throughout the return voyage numerous small—and occasional larger—mechanical and electrical faults occurred. These were generally as a consequence of the fire damage, or because the ship had been without heating for a considerable period. We limped into Hobart, with tug assistance for the final leg, at 3:30am on 31 July.

By the same time of year in 1999, the ship had been repaired and we were ready for a second attempt. This time the 62 scientists and technicians on the ship spent nearly six weeks in July and August within the sea ice zone, investigating oceanographic and glaciological processes that are related to global climate, and also undertaking biological studies. The oceanographers measured the ocean properties and structure, from top to bottom, at 25 stations around the core of the polynya. Each station was measured three times to determine changes in the ocean as new ice formed on the surface. They also deployed recording instruments on sub-surface moorings to continue measurements over the next six months. Meanwhile the glaciologists estimated the ice and snow cover thickness and structure, and their change with time, from observations from helicopters and the ship and from measurements at 42 sites on the ice. We also deployed 21 satellite-tracked buoys to record the rate and pattern of ice transport across and out of the polynya. Three of these buoys also provided meteorological data to supplement the continuous meteorological measurements made onboard the ship, plus daily observations of the atmosphere at height using helium-filled balloons launched from the ship.

The combined oceanographic and glaciological observations showed that ice formation rates in the MGP were around five to eight centimetres per day. Brine rejected during this growth increased the winter salinity over the continental shelf and played a significant role in bottom water formation that drives vertical ocean circulation.

Our multi-year observations over the East Antarctic sea ice zone, between 20°E and 160°E, showed the pack to be highly mobile with an average drift speed of nearly 20 kilometres per day. However, the drift is highly variable on a daily basis. The generally divergent nature of this movement constantly opens new areas of open water and nearly 50 percent of the total ice mass forms from rapid freezing in these leads. Only about 40 percent of the ice within the pack forms from slow basal freezing, the usual process for land-fast ice growth. The remaining 10 percent forms when a heavy snow load swamps the ice surface, leading to rapid freezing of the seawater saturated snow near the surface. The area-weighted average thickness of the ice in this region is less than one metre throughout the year.

The rapid decline of sea ice in the Arctic has been one of the most significant recent climate-related changes. Over 40 years of satellite observations, the Arctic sea ice extent has decreased in all seasons, with the largest decrease of almost 13 percent each decade occurring in summer. At the same time, the average thickness of Arctic sea ice has thinned, mostly due to a decrease in the proportion of thick multi-year ice that survives into the summer. Arctic surface air temperature has increased by more than double the global average over the last two decades, and feedback from the loss of sea ice has amplified this warming. However, the Antarctic sea ice cover has shown no significant trend in extent over the same period, and there is insufficient accurate data to assess any changes in its thickness.

But the Arctic Ocean and Southern Ocean sea ice covers have quite different characteristics. Most Arctic sea ice lies north of 60°N and is surrounded by land to the south with openings to the Atlantic and Pacific oceans. Some ice remains trapped within the Arctic Basin for several seasons, thickening by basal freezing and deformation (ridging and rafting). In contrast, Antarctic sea ice only occurs north of 78°S, north of the ice sheet and ice shelves. It is unconstrained to the open ocean in the north, and is largely seasonal, with only a small fraction surviving past the summer minimum in February. Unconstrained by land boundaries, the latitudinal extent of the Antarctic sea ice cover is highly variable and driven by winds.

Some response of Antarctic sea ice to changes in atmospheric and oceanic temperature, and to changes in wind speed and patterns is inevitable. But trends in the ice coverage may be due to trends in the surface wind, and without better ice thickness and ice volume estimates, it is difficult to characterise how Antarctic sea ice cover will respond to changes in climate.

Glaciers on sub-Antarctic Heard Island

My first foray into studying the terrestrial cryosphere (land ice) was in the summer of 1970-1971 when I participated in an expedition of Terres Australes et Antarctiques Françaises (TAAF) to heavily glacierised sub-Antarctic Heard Island in the Indian Ocean at 53°S, 73.5°E. The French were investigating propagation of Very Low Frequency radio waves by making simultaneous observations at a number of sites, including Heard Island, on the same longitude. Since the island was Australian Territory, they invited Australian participation. Heard Island had a total area of glaciers of about 280 square kilometres at this time, but the total area had been steadily retreating since at least 1947. However, there were almost no observations of the physical characteristics of the glaciers—such as thickness, movement and the processes of snow accumulation and ice melt. My then Antarctic Division supervisor, the internationally eminent mathematical glaciologist and climate scientist, Bill Budd, proposed a program of such measurements on a Heard glacier, and to use these measurements to also test an empirical scheme he had developed of estimating the dynamics of unmeasured glaciers. I was delighted to be offered the opportunity to make these measurements. I joined four other Australians and an eight-man French party, and we travelled to Heard from Mauritius via Reunion and Iles de Kerguelen on the French vessel Gallieni.

We arrived at Atlas Cove, the site of the old Australian station on the northwest of the island, on 25 January 1971. After first chasing elephant seals out, we established our living quarters in the old buildings and set up the French experiment in some new portable buildings. I set to work with the help of the other Australians on a survey of the Vahsel Glacier, about five kilometres south of our camp.

Bamboo canes were placed in the ice to measure snow accumulation and melt rates. Their positions were surveyed by theodolite resection to nearby rock peaks (when visible in the frequent mist and cloud), the surface profile across and along the glacier were measured by optical levelling, and a network of surface gravity measurements were made to estimate the ice thickness. Because ice has a density of only about a third of that of rock, the gravitational force on a glacier surface is less than it would be for a rock surface at the same elevation and, after numerous corrections for elevation and the surrounding terrain, these measurements can be used to estimate the thickness of the underlying ice.

On 7 February, three of the Australian party started an anticlockwise walk completely around the island to survey changes to the glaciers and to census the fur seal and king penguin populations. Heard Island is dominated by a central, active volcanic peak with an elevation of 2,745 metres. More than 20 glaciers spill from near its summit to the sea, making travel very difficult, particularly on the western, windward side of the island where snowfall is very high and the glaciers are very dynamic.

I continued work on the Vahsel Glacier while the one Australian physicist in our team worked with the French physicists. But six days later we were surprised with the return of two of the circumnavigation party to the base. The third member of their party, Ian Holmes, had fallen badly while jumping a crevasse and broken his leg. This occurred on February 10 on the very deformed Gotley Glacier, nearly 20 kilometres south of Atlas Cove. His leg was temporarily splinted and he was left with supplies in a tent on the glacier surface while the others returned to seek help, continuing on the longer but easier anticlockwise route. It took them three days to get back, during which time the tent collapsed and left Ian lying poorly protected on the melting glacier surface.

To rescue Ian, it was necessary to recall MV Nella Dan, with helicopters, from its program at Mawson station, about 1,500 kilometres south of Heard Island. The ship had mechanical problems on the way but reached Heard on 21 February. Ian was winched to safety, after 11 days alone on the surface of the Gotley Glacier, on an extremely rare day when weather conditions allowed flying a helicopter under visible conditions over the glacier.

I completed the resurvey of the Vahsel Glacier to determine ice movement and surface mass balance in the remaining time before we were retrieved by Gallieni on 8 March. The results of the survey showed that the Vahsel, which is about nine kilometres long, is about 60 metres thick and moving at up to 270 metres per year near the equilibrium line (which separates that part of the glacier that annually gains mass from snowfall, from that which loses mass by melt). These results validated Bill Budd's empirical estimation scheme. Our survey, together with subsequent satellite imagery, also showed that the Vahsel, named after the second officer on the Gauss during Erich von Drygalski's 1901-1903 Antarctic expedition which had visited Heard Island, was actually two separate glaciers. This was recognised with the second part renamed Allison Glacier.

Until recently, glaciers on the western side of Heard Island, like the Vahsel, were tidewater glaciers. They terminated at the coast and part of their mass loss was from calving of small pieces of ice directly into the ocean. With climatic warming, their initial mass loss showed as a decrease in surface elevation, rather than as a change in area, which is easier to detect from simple visible observations. However sometime before 2010, satellite imagery showed that the Vahsel and Allison glaciers started retreating inland. A recent (November 2022) Sentinel-2 satellite image, which showed that the Heard Island stratovolcano was erupting, also showed that the retreat inland had extended to other west coast glaciers. The front of the Allison Glacier is now about 350 metres from the coast: climate change certainly becomes more 'personal' when it impacts your eponymous glacier.

The glaciers on the eastern rain-shadow side of Heard Island have long terminated on land. They have shown rapid retreat since 1947, and at an accelerating rate since the mid-1980s. In the early 2000s, I was involved (but not in the field measurements) in a study of one of these, the Brown Glacier. It decreased in area from 6.2 square kilometres in 1947 to 4.4 square kilometres in 2004. Between December 2000 and December 2003, GPS surveying showed that its surface elevation had decreased by more than six metres, more than double the 57-year average rate of decrease. Meteorological observations, which are not continuous at Heard Island, indicate a 0.9 degrees Celsius warming over the same time.

Equatorial glaciers of New Guinea

Two years later I had the opportunity to participate in a survey of small mountain glaciers in a far different regime. This was as the field glaciologist on the second Australian Universities Carstensz Glaciers Expedition (CGE) in early 1973.

The Carstensz Glaciers, on the tropical island of New Guinea, were first sighted in 1623 by the Dutch explorer Jan Carstenszoon, when he sailed through Torres Strait, between New Guinea and Australia. His sighting was unverified by Europeans until the Dutch assumed control of the western half of New Guinea, more than two and a half centuries later. A large British Ornithological Union Expedition made an unsuccessful attempt to reach the high peaks from the south coast in 1909-11. An even larger expedition of 224 men led by AFR Wollaston in 1912 spent three days at over 3,000 metres altitude, and just reached the edge of the ice. The far smaller aircraft-supported expedition led by AH Colijn in 1936 was much more successful. This included the young petroleum geologist, JJ Dozy, who marked the fronts of the glaciers with cairns and discovered very high-grade copper ore in the region. This expedition established approximate heights of the main peaks and took oblique aerial photos of the glaciers. Trimetrogon aerial photographs (with one vertical and two oblique cameras) taken by the US Air Force in 1942 indicated significant retreat of the glaciers.

After the war, these high peaks came to the attention of climbers. A New Zealand expedition followed native trading tracks from the north in 1961 but did not reach the ice fields because of a failed airdrop of supplies. The highest peak, then called Carstensz Pyramid, was first climbed in February 1962 by Heinrich Harrer, who in 1938 had been the first to climb the north-wall of the Eiger in Switzerland. Harrer recorded continuing retreat of the Carstensz glaciers.

West Papua came under Indonesian administration in 1963, and in 1969 it controversially became a sovereign territory of Indonesia. In January 1970, after considerable exploratory work, Freeport Indonesia Inc. commenced construction of a commercial copper mine at Ertsberg, only about three kilometres directly from the highest peak and about a 10-kilometre trek to the main glacier fronts. An airfield, road to the south coast, aerial tramway and other infrastructure were built and the first ore from the mine was processed in October 1972.

These facilities provided a means of access for scientific investigation of the glaciers. Freeport Indonesia Inc. provided support, including a helicopter cargo lift, for the first CGE, a party of six scientific personnel, who occupied a base camp at 4,250m elevation from December 1971 until the end of February 1972. This camp was within 500 metres of the tongue of the Meren Glacier, one of the two major valley glaciers, the other being the Carstensz Glacier. CGE-1 began a detailed trigonometric survey, linked to the mine survey which had been extended from the coast. This was then used for photogrammetric control of mapping using the 1942 USAF photos. They established networks of canes on both valley glaciers for measurement of ice movement and snow accumulation and melt and made gravity observations to estimate ice thickness. Meteorological and hydrological stations were established with recording instruments and surveys were made of flora, fauna and landforms.

Logistic support from Freeport was unavailable to deploy CGE-2 the next year, so our five-person scientific team had to reach the glaciers overland from the north. We flew via Port Moresby in Papua New Guinea to Madang on New Year's Day 1973. Then, the next day, from Madang to Jayapura on what was probably one of the last internationally commercial air connections still operated with a DC3 aircraft. In Jayapura it was necessary to arrange air transport and travel permits for the highlands. Here we were also joined by Sam Mustamou, an Indonesian student from Cenderawasih University, who became a member of the expedition team.

We met the civil vice-governor of Irian Jaya, who I was impressed to find was a native Papuan. But ultimately, our highland travel permit had to be obtained from the Indonesian military vice-governor, whose headquarters were in caves outside Jayapura that had been used by General MacArthur during WWII. We were advised that site protocol was that no one below the rank of colonel should wear sunglasses.

Expedition leader Jim Peterson and all other CGE-2 members, except Sam and I, flew to the highland village of Ilaga on 5 January in a Safari Air Norman Islander aircraft. Safari Air was chartering to Newmont Indonesia Inc. but able to divert and add this extra flight. Ilaga, at an elevation of 2,331 metres on the Ilu River, had two air strips and two missions on either side of the river. The western side had a Catholic mission serving Amume Damal people and had an airstrip operated by Associated Missionaries Aviation (AMA). On the other side was a protestant mission serving Western Dani people, the government post and an airstrip, which was also used by scheduled and charter flights, operated by the Combined Aviation Missionary Alliance (CAMA).

I remained in Jayapura to purchase extra rice and tinned fish for the porters for our trek in and to arrange the uplift of the rest of our equipment and stores. The latter was not initially easy. The next scheduled flight to Ilaga was not for about another two weeks and the aircraft of a couple of other charter companies were too large for either airstrip at Ilaga. But on 8 January a planned flight by Safari Air for Newmont could be diverted because equipment had not cleared customs in time. Another 550kg of our equipment was hence transported to Ilaga by the Islander.

Finally, after an evening visit to Father Dykman at the Jayapura Catholic Diocese HQ, I was able to arrange an AMA flight. The Catholic mission at Ilaga was run by a long-serving Dutchman, Father Schins, who lived much the same lifestyle as his parishioners. Father Dykman had received a request from Schins that I might bring two things: a bottle of whisky and a sub-machine gun to 'wipe out the protestant bastards'. Dykman recommended that I take only the first gift. On the morning of the next day, January 10, Sam, I and the remaining 180kg of cargo reached Ilaga in an AMA Cessna. We shared a coffee and a 'wee dram' with Father Schins and engaged porters to transfer the gear the short distance to the Protestant side of the river.

Ilaga is about a 70-kilometre trek east of the glaciers. Jim had employed 33 porters to carry our equipment and stores and we were also assigned a local native policeman, Max Katau, who was equipped with an ancient .303 rifle for our protection on the trek. We repacked everything into 20kg loads and were away early the next morning with the Dani porters, led by local preacher Georit Wanembo, plus about 20 family and friends. The Dani carried the loads on their heads, were bare-footed and wore nothing other than penis sheafs and a few body decorations. We crossed the Ilu River and climbed over 1,000 metres that day, up the north-facing scarp of the Zengillorong Plateau, through mountain forest, over streams and fallen logs.

The second day we continued upwards through mossy cloud forest with high trees and a thick canopy that persisted to near the edge of the plateau where it gave way to sub-alpine forest and shrubby grasslands. The native hut at the second night's stop had burned down, it was raining heavily, and we all spent a miserable night. Several porters deserted but were replaced with recruits from among the accompanying family and friends. I recruited a personal assistant, Willem Murip, the 10-year-old son of one of the porters. He carried a 1kg load on his head and became my best mate after I gave him some cast off, and far oversized, Antarctic-clothing and some beads. He completed the whole trip and frequently egged me on whenever I started to falter.

Travelling became easier on the karst plateau, although the rains always came in the afternoon and the nights were generally uncomfortable. I shared a tent with Max and Sam throughout the trek and taught them all the songs I knew from the Fabian Society (socialist) song book. Probably not wise given the politics of Indonesia at the time! Max only used his rifle once: from inside the tent in the middle of the night, ostensibly to scare off a large black dog. But dogs, particularly black ones, are also considered good for eating. On the fifth day we got a first glimpse of the distant ice.

On day six we reached Lake Larson at the base of the climb over a high pass to the ice. The next morning the near-naked porters had a meeting to decide whether they should take loads over the pass or not. They split into two groups—those willing to go and those not. Georit and Sam called a prayer session for the Holy Spirit to decide. Fortunately, the Holy Spirit was on our side. Most of the porters agreed to continue after a breakfast of fish and rice, and after Jim provided them somewhat warmer clothing he had bought along.

We finally started the gruelling climb at 10:30am, up through New Zealand Pass and down to the base camp site in the Meren valley. After re-establishing the camp and retaining Georit as camp assistant, we were able to commence our scientific program. The CGE-2 glaciology program involved a resurvey of ice stakes to determine net snow accumulation and ablation and ice movement over the 12 months since the first expedition. Five boreholes to 10 metres depth were cored on the Meren Glacier and three on the Carstensz Glacier to measure temperature profiles within the ice. Additional temperatures were measured in deep crevasses. Ice samples were collected from the boreholes for subsequent stable isotope analysis. The snouts of both glaciers had receded nearly 10 metres further since the previous year. Twice-daily meteorological observations were made at the base and long-term meteorological and hydrological recorders, which had operated for more than nine months, were reactivated.

The surveyor, Ted Anderson, continued strengthening the

existing survey, including the ground control points to enable photogrammetric plotting from the USAF photos. He and Jim climbed the Carstensz Pyramid and placed a marker pole on the summit, enabling triangulation to accurately determine its height as 4,884m. This was somewhat lower than earlier crude estimates, but the Pyramid is the highest peak between the Himalayas and the Andes. The snow-covered peak Ngga Pulu, at the top of the Meren Glacier, had a height of 4,860m, now decreased with ice loss.

Biological investigations were made of the remarkable cryo-vegetation colonies growing on the snow and ice, including the prolific algal mats on the glaciers. Sediment cores were collected for investigation of vegetation histories and glacial deposits in the high valleys were mapped to determine the Pleistocene glacial history of the Kemabu Plateau, north of the Jaya Massif. The work program was completed by late February.

The two multi-disciplinary CGEs resulted in the first accurate map and glacier survey of the region. Measurements of the glacier processes were used in a simple numerical glacier model to show that the glacier retreat since 1850 was compatible with an atmospheric warming of about 0.6°C per century, similar to what had been observed at lower elevations in the region. Continued warming at this rate would see both valley glaciers disappear in the first half of the 21st Century. The geomorphology and palaeobotany revealed a complex pattern of ice movement over the past 15,000 years, including evidence for glacier changes over the past 30,000 years. The biology included an ecological survey of the vegetation and the remarkable algal communities growing on the ice, revealing how the flora is specialised for the alpine environment. And preliminary archaeological evidence showed that man has been visiting the high mountains for the last 5,500 years.

Retreat of the glaciers on both Heard Island and the high New Guinea mountains is clearly evident and related to a warming climate. In New Guinea it has been ongoing since about 1850, and on Heard since at least 1945. In both cases, glacier retreat has accelerated in the 21st century as the average rate of warming has increased. The Meren Glacier disappeared completely in about 2000 and the Carstensz will shortly follow, somewhat earlier than our model projections which were based on the rate of warming in the 1970s. Globally, the rate of glacier wastage increased in the 1980s and accelerated further in the early 21st century. Meltwater eventually flows from the glacier to the ocean and a net loss of global glacier mass contributes to a rise in sea-level. The estimated contribution of glacier retreat to sea level for 2006-2016, excluding the glaciers peripheral to the ice sheets, is 0.6 mm/yr. The current best estimate of the total number of glaciers on Earth is over 215,000, and these contain the equivalent about 0.3 metres of sea level (perhaps a little less after subtracting water retained in lakes of the subglacial topography, or stored in closed basins). A rise of this magnitude would have significant impact on human societies due to the concentration of communities and infrastructure in coastal regions. This potential sea level rise is however small compared to what could be contributed from the Greenland and Antarctic Ice Sheets.

The polar ice sheets

Within the past 125,000 years, variations in Earth's climate have resulted in global sea levels fluctuating from 130 to 140 metres lower than present-day to six to nine metres higher. These changes have been primarily driven by the growth and decay of ice sheets in the Northern Hemisphere. The only remnant ice sheets today are in Greenland and Antarctica. Respectively, they contain the equivalent of 7.4 metres and 58.3 metres of sea level. These ice sheets gain mass from snowfall which is slowly metamorphosed to ice and transported from the high interior to the coast by ice flow. This is a very slow process, and there is likely some ice in Antarctica that first fell as snow over one million years ago. Ice is lost at the coast by surface melt and run-off (not very important in Antarctica) or by direct discharge into the ocean. In Antarctica, most of the discharge is from faster moving ice streams which push floating ice tongues or ice shelves out onto the ocean. Loss from these occurs from iceberg calving or from basal melt where they are in contact with the ocean. Any imbalance between the snow-mass input and the ice-mass discharge will impact global sea level. Although the pace of change may be very slow, once in place climatically induced imbalance can be inexorable and persist for hundreds or thousands of years.

Understanding the state of the Antarctic mass balance has been a 'holy grail' of Antarctic glaciology since at least Wright and Priestly considered it in their glaciological report of the scientific results from R.F. Scott's (final) Terra Nova expedition in 1911. There have been numerous estimates since, but based on sparse field observations and providing no definitive answer because they are estimating a very small difference between two very large numbers and with large uncertainties.

Bill Budd, my 1970s-supervisor, was a pioneer in developing the theoretical basis of ice sheet dynamics and in ice sheet modelling. Both were based on his own field studies in the early 1960s on the Law Dome ice cap (110°E) and the Amery Ice Shelf

(70°E). Bill conceived and supervised a program that involved a very small party of four men, overwintering on the Amery Ice Shelf in 1968, undertaking a detailed survey of its characteristics and dynamics and drilling an ice core. Following the success of this, the next step was to determine the processes and the mass of ice flow into the Amery from the grounded ice sheet via the enormous Lambert Glacier system. Bill proposed a program to address this and assigned me to undertake the field measurements.

This was undertaken during the Austral summer of 1971/72 as part of an air-supported, multi-disciplinary field programme in the southern Prince Charles Mountains (SPCM). My colleague, Mick Skinner, and I established eleven ice-movement markers at about 2,000m surface elevation around a perimeter inland of the mountains, covering all the major tributary ice streams that fed into the Amery. We were helicoptered to each site where we camped for two to three days to survey a precise location by theodolite and electronic distance measurements to parties on mountain peaks up to 100 kilometres or more away.

This first ice sheet expedition was nearly my last! At the fifth site, while waiting to be transported to the next, I looked for a small crevasse in which I could measure the temperature at 10 metres depth (which approximates the annual average temperature). Carelessly un-roped and unable to find a suitable crevasse, I impatiently stomped up a small rise—and found one large enough to take me. I broke through a snow bridge and fell about 3 metres to an internal snow bridge. This broke my fall, but it sloped downwards, and I slid another 2-3 metres before coming to a halt on a slight lip at the edge of dark blue nothingness. Mick went back to the tent to get a rope and caving ladder. I was warmly dressed and used a pocket-knife to carve foot-holes in the crevasse wall so that I could chimney myself in—I didn't trust the bridge to hold. But it seemed to take ages for Mick to return and I was not sure that he had not got into trouble himself. I clambered out shortly before the helicopters arrived—so no one, other than Mick, knew what had happened (you did not have to report stupid near-misses in those days). I have remained 'crevasse wary' ever since.

I went back to the SPCM in the summer of 1973/74 to remeasure the sites and determine the ice movement. These velocity measurements, combined with aerial radio echo-sounding measurements of ice thickness, enabled the total flux of ice across the perimeter to be estimated. The flux about 500 kilometres downstream, through the Amery Ice Shelf, was considerably less than that across the 2,000-metre contour indicating, after correcting for snowfall and surface melt, that the section of the ice sheet between must be gaining mass and increasing in elevation.

The scientific paper that resulted from that program has been one of my most frequently cited—but this conclusion turned out to be wrong! From the data available in the 1970s it appeared that the ice started to float a short distance downstream from the Amery flux section. But satellite data that became available in the 1990s showed that the Amery Ice Shelf actually started floating 240 kilometres upstream of the previously reported position. The excess ice from the 1970s estimate was being lost as melt from the base of the ice shelf.

I undertook similar field projects to estimate the ice sheet mass balance in the coastal region of Enderby Land and Kemp Land (between about 50°E and 60°E longitude) in the summers of 1977/78 and 1978/79. The latter season included a program of flying a radio echo sounder in a single-engine ski-equipped aircraft to map the ice thickness. One memorable incident during that season occurred when we landed to refuel from a tractor train that was returning from the field camp to Mawson station. While there the weather deteriorated and rapidly approached blizzard conditions. To protect the aircraft from wind damage, we firmly tied it in the lee of two large tractors before retiring to a sled-mounted caravan. When, a couple of days later, the blizzard had passed, we found the aircraft well and truly buried and snow-filled. It took several hours of hand digging before we could become airborne again.

Technological advances have made remote ice sheet measurements increasingly easier and more accurate. During the 1977/78 and 1978/79 programs we used the Transit satellite (doppler) navigation system to determine the location of ice movement stations. After several days of continuous observations, a location with an accuracy of several metres could be obtained without reference to a fixed point such as a mountain. I went back to the PCM to make more measurements on the Lambert Glacier in 1988/89, this time using GPS receivers. In those days a geodetic GPS receiver was the size of a small suitcase and, because the satellite constellation was not complete, there were only a few hours in each day when there were enough satellites visible to give a good 3-D location. A navigational GPS unit was also used to map the flight lines of radio echo-sounding flights over the Lambert Glacier, although with the same satellite availability constraints. These ice thickness flights were made both from a twin-engine aircraft brought to Antarctica by the Australian businessman and adventurer, Dick Smith, and from a helicopter.

However, the greatest advances in understanding the

Antarctic ice sheet mass balance have come with satellite remote sensing. Since 1992, three different and independent satellite-based methods have become available to estimate changes in the ice mass of Greenland and Antarctica. These use satellite measurements of surface elevation (both radar and laser); satellite measurements of the gravitational field; and satellite synthetic aperture radar (which can map surface ice velocity). These give broadly similar conclusions, indicating that Antarctic Peninsula and West Antarctic Ice Sheet have both lost mass since 1992 and that the rate of their combined loss increased after about 2006. The combined loss contributed 0.17 mm/year of sea-level rise from 1992 to the end of 2006, increasing to 0.51 mm/year SLE from 2007 to the end of 2016. Changes to the larger East Antarctic Ice Sheet are less certain. It remained near in-balance, with large inter-annual variability due mostly to fluctuations in snowfall, but there are recent indications that some regions of East Antarctica may also be starting to experience significant loss.

The satellite techniques are now giving mass balance estimates over the entire ice sheet that could never be previously made with isolated field programs. However, interpretation and even knowing what to measure with satellite technologies, has necessarily been built on earlier field studies that define the boundary conditions for measurement and identify the processes that are important. Not all the processes that might affect the response of the ice sheet to global warming are yet understood.

Concluding remarks

This recollection, based largely on memory and personal diaries, has undoubtedly over-emphasised my own involvement. But all the projects described have very much depended on collaborators and supporters. There have always been many colleagues involved in the conception, planning and preparation of projects; in undertaking and supporting the field programs; and in analysing, interpreting and drawing conclusions from the results.

Collaboration is necessary not only between individual scientists, but between many institutions and many nations. Many of the interdisciplinary science questions we are addressing today, particularly those related to climate change, are far too big for one individual, for one institution or for even one nation. International organisations, both governmental and non-governmental, play a significant role in these international collaborations. These organisations do not directly conduct research programs, but they bring together experts from many nations to promote, plan and coordinate large-scale cooperative programs.

My perception of the impacts of climate change on the cryosphere has changed over the decades. Early direct field observations of changes to small glaciers, such as those on Heard Island and New Guinea, could be associated with changes in local temperature and precipitation. But estimated changes to the Antarctic Ice sheet were then very uncertain; made over vast areas based on limited, and biased, field observations. Modern satellite technology has changed that. We now can measure with reasonable accuracy what is happening overall to the large, but uniform and conterminous, ice sheets. But while we may know precisely what is happening to one small glacier, we have an incomplete picture of the cumulative changes from many small glaciers in many different environments. I expect that evolving technology will resolve this.

For our future, sea level rise is a major consequence of climate warming. Presently, global mean sea level is rising at its fastest rate in the past 6000 years. Since 1992, melting glaciers and ice sheets have contributed about 50 percent of the total sea-level rise and are expected to continue to contribute this proportion for the rest of the 21st century. Projections of future sea-level rise, required for adaption planning, have large uncertainties, partly because of incomplete understanding of all the processes of ice sheet dynamics. But we do know many ice masses worldwide are out of equilibrium with the present climate and will continue to shrink and add to sea level even if global temperature stabilises.

BETWEEN A ROCK AND A HOT PLACE

BY

LINDA BROOME

Dr Linda Broome *was born in Orbost, Victoria, Australia, a small town on the banks of the Snowy River, on 10 May 1955. She grew up at Noorinbee, an isolated farming community in the Cann Valley bounded by extensive forests in far East Gippsland. Her empathy for fauna and flora developed from natural curiosity and her father's tendency to rescue any injured birds found on the roads. She received her first bird identification book at age 10 and spent her free time as a teenager riding her horse through the bush observing the natural history of the area. Linda attended the University of New England from 1973-76 for a degree in Natural Resources, specialising in wildlife management. She worked as a research assistant in the School of Natural Resources, UNE, from 1977-1979, with some research forays to Papua New Guinea and New Britain, then undertook a PhD in Ecology/Biology at Utah State University from 1980-85. Linda returned to Australia for a research project on the mountain pygmy-possum in 1986. She began work with the New South Wales National Parks and Wildlife Service in 1995 and is still employed in what is now the Department of Planning and Environment.*

January 1986

I had just driven from Sydney to the Snowy Mountains in far south New South Wales and was easing the vehicle steadily up the newly made steep, dirt track to Mt Blue Cow. I had flown in from Utah the previous day, collected a vehicle from National Parks and Wildlife Service (NPWS) Head Office in Kent Street, negotiated Sydney traffic heart-in-mouth, and crossed the Harbour Bridge twice after staying overnight in Sydney. No mean feat for a city-phobic in pre-GPS days, street directory in lap, praying for a red light so I could glance down, after living in the USA and driving on the right-hand side of the road for the past five years.

Breathing a sigh of relief and taking a deep breath of fresh, Eucalypt-laden mountain air after my five-hour drive from the city, I rounded the last bend, looked up at the mountain peak towering above me and immediately swore, deriding myself with thoughts of, 'Hells teeth, now what have I let myself in for?!'

I had just completed my PhD research at Utah State University on small mammals, radio-tracking deer mice in Wyoming on gently sloping hills and broad swales. There it was easy walking:

during summer on sandy soils between the sage brush and up to slightly higher vantage points to gain reception with the tracking antenna, or traipsing across the snow above them during winter on snowshoes. Granted, winter at minus 40 degrees Celsius, not counting wind chill, had its challenges. Not least of which was avoiding stepping outside the van in the Kemmerer, Wyoming, trailer park (town motto: 'Buckle of the Overthrust belt') with wet hair causing it to snap freeze and break!

As I gazed at that steep, boulder-strewn slope above me, I realised the strong likelihood of more than my hair breaking in these Australian mountains. The slope was covered with shrubs at its lower elevations, cascades of boulders cloaked the edges and an endemic conifer, the mountain plum-pine, low-growing and hugging the surface of the rocks for warmth and wind protection, had the potential to catch the foot of the unwary.

My task was to live-trap and radio-track a newly discovered population of mountain pygmy-possums that were known to occur on the site and to suggest means by which the effects of the developing ski resort might be ameliorated. Plans for the resort were well underway, but very little was known of the possums' numbers, their movements, ecology, significance of the site in relation to the whole population, or the impacts the ski resort might have on them.

Not to be overly daunted, I thought back to the moment at the start of my PhD venture when I discovered there was a thing in the cold climes of the USA called a 'field season', whereby most of the graduate students conducted their research from May to September and spent the winter months indoors.

A little puzzled, I had responded, 'But in Australia we do field work all year round.'

The rejoinder, a dry 'go ahead', was met with hands on hips and all the confidence of a 20-something declaring to my supervisor and all in earshot, 'I will'.

And I did. Nobody had radio-tracked mice during winter in Wyoming previously and they probably haven't since!

As my quadriceps and sure-footedness grew, so did my love of that mountain and its inhabitants. In the early morning light, when checking traps set for the possums, my eye was drawn east across the rolling plains of the Monaro tableland and a little north to the Brindabella Ranges of the ACT, shrouds of mist rising from the valleys with banks of fog low on the horizon indicating the coastal lowlands beyond. The trap-line ascent to the peak of Mt Blue Cow, at 1,984m, afforded a breath-taking view to the west across to the Main Range and Mount Kosciuszko, the highest peak on the Australian mainland (2,228m). Summer dusks on the peak were accompanied by a low hum from the depths of the boulders beneath, rising to a crescendo of a million tiny wings as the migratory bogong moths took their evening flight.

Bogong moths undertake a remarkable annual spring journey of up to 1,000 kilometres to escape the heat of the lowlands, spending the summer months sheltering in the cool rock crevices of the mountains, before making their way back in autumn to breed. There were magic nights as I followed the signals of the radio-collared possums with the stars of the Milky Way blazing overhead and bats swooping for the bogong moths. Some male possums took me from low on the mountain, across the peak to Guthega and back in the early morning—a three-kilometre round trip in the dark with a bold, mate-seeking, gene-spreading, 35-gram possum. Even the females who nested at the lower elevations would travel up to 500 metres to the peak to feed on the bogong moths that concentrated there in the hottest part of summer. Sometimes they headed out several times in a single night, their rapidly growing young left in the nest as they were too large to be carried in the pouch. The deluxe habitat was on the peak, with females there moving less than 100 metres in summer and winter.

Winter afternoons initially saw me skiing in on cross-country skis, though later on an energy-saving, hand-warming, noisy, but highly entertaining snow mobile. Seeking the nest sites of possums and checking them regularly throughout the night, some nights turned to blizzards with reem ice coating my jacket and the branches of snow gums snapping under the weight of ice accumulation with the loud report of gunfire. The wind, picking up through the night with all the strength of the Roaring Forties behind it, once carried me to the edge of a cornice under which the possums were safely ensconced, and then over in a tumble of skis and radio tracking gear. Other nights were calm, with the crunching cold of skis or snowshoes as I traversed the mountain, moonlight reflecting from the snow so brightly I had no need of my head lamp to write notes.

This was repeated two years later, with the night work taken over by data loggers at nest sites recording body temperatures of possums carrying thermal sensors. Skiing across the mountain with radio antenna in hand to locate individuals that occasionally moved position, I heard wisecracks from clients on the newly opened chairlift above me, 'Lost your TV, love?'

In my mind the grandeur and peace of the mountain was now diminished. Though I discovered, unlike the deer mice in Wyoming which remained active under the snow throughout winter, that mountain pygmy-possums hibernate for up to seven months underneath their deep, doona covers of snow.

27 May 2010

With my new PhD student, Haijing Shi from China, I had just hiked the kilometre-long boulder field down the western fall of Mt Kosciuszko, across the valley, ascending past Muellers Peak, on to Mt Townsend, down to the treeline on the steep western side, distributing data loggers on our way, and back to the vehicle at Rawsons Pass. Hayley Bates, my other new student, had prepared dinner but Haijing quite literally fell asleep with her face in her plate!

I had completed the three-year intensive study on Mt Blue Cow, with three additional sites added for comparison, at the end of 1989. In the intervening years, with the help of many wonderful volunteers, the possum and bogong moth numbers at these sites had been monitored annually. My baby daughter Rose was born in 1991, graduating from belly to backpack, to hip, to being sent down holes to retrieve traps, to data recording and possum wrangling as she grew.

We had also mapped, trapped and assessed every boulder field on the Main Range from Thredbo to Gungartan Pass, 30 kilometres to the north. Our research indicated that Mt Blue Cow and The Paralyser, on the western boundary of Perisher Ski Resort, supported the largest populations of mountain pygmy-possums in NSW, along with another ski resort, Charlotte Pass, and Mt Kosciuszko and Mt Townsend.

The highest densities of possums occurred in deep boulder fields at high elevations, which sheltered large numbers of bogong moths throughout the season. The lowest elevation we found the possums was 1,600m, although later we found a colony at 1,500m in a pile of rocks on the Snowy River near Guthega, excavated during the construction of a tunnel for the Snowy

Mountains Hydro Electric Scheme. Abundance was also related to a diverse shrub and forb cover surrounding the boulder fields, and the percentage cover of snow between July and September. No surprise then that the best places for skiing coincided with the best habitats for the mountain pygmy-possum.

My Honours student, Bec Gibson, had spent months staring into a microscope identifying the remains of plant and insect material in the scats of the possums we had collected over the years. Bec determined the seeds of the native conifer, the mountain plum-pine, were as highly nutritious and important to the possums' diet as the fat-rich bogong moths. However, a range of other seeds, fruits and arthropods from the surrounding shrub lands are included in their diet, the amounts varying by year, season and elevation. Bec now wears glasses!

The question I posed to Haijing and Hayley was, 'What limits the lower elevation of distribution of the mountain pygmy-possum?'

Haijing was looking at temperature regimes in the boulder fields at different elevations, aspect and depths, with and without plum-pine cover, while Hayley was looking at a range of other variables including food and water availability, predators and competitors. Our first mission was to locate a range of boulder fields below the current known 1,600m elevation limit that did not have possums.

The next day Haijing and I headed to the northern part of Kosciuszko National Park where a botanist colleague had told me there were some basalt boulder fields at 1,600m and below. We drove from the most northern known possum site at Gungartan, through a couple of tricky river crossings and past the skulking bulk of Mt Jagungal.

I had previously trapped on Jagungal in 1989 after a helicopter drop-off, with no possums located. This was not surprising, due to the limited extent of only shallow boulder fields on that monolith. The boulder fields in the southern part of Kosciuszko National Park and in the far north at the Bogong Peaks are granitic, with giant tors in some places as large as trucks. The ones favoured by possums have small rock sizes many layers deep, with small gaps in which they can hide and where temperatures remain cool and constant. The north central part of the park is an ancient basalt flow, with boulder fields of small to medium sized rocks erupting along creek lines and mountain terraces. We located a succession of these, ending up at Rough Creek, north of Round Mountain. The boulder fields at Rough Creek are extensive, ranging from 1,400m to 1,660m elevation and were fringed with mountain plum-pines.

I looked at Haijing and mused, 'You know, there could be possums here, it will be interesting to trap next spring.'

I had previously asked our botanist to collect some of the dropped leaves and seed cases from under these plum-pines but had not had time to examine them. A plum-pine seed eaten by a pygmy-possum can be distinguished from one eaten by a bush rat because the possum holds the seed in one paw, placing it sideways in its mouth and breaking it open transversely with a specialised large pre-molar tooth with a saw-like cutting edge. Rodents will hold the seed in both paws and bite down on it with the incisors, cutting it lengthwise.

With the first winter snow lightly dusting the rocks in mid-June, we finished placing data loggers along our route and in an additional impressive boulder stream crossed by the road near Cabramurra. We returned by Happy Jacks Valley, a steep, narrow riverine gorge at 1,200m flanked by boulder screes and overlooked by 1,687m Boltons Hill. As we rounded the corner and looked across to Boltons Hill, I could see extensive basalt boulder fields, enhanced by the sprinkling of snow, descending in waves to the creek on its steep southern flank.

With some trepidation, I thought, 'We are going to have to trap there.'

The following spring, busy with the possum monitoring at the southern sites, I received a text message with a photo attached from a colleague who had started a survey of Happy Jacks Valley prior to the rehabilitation works of the old hydro scheme spoil dumps in the area: 'I think I have trapped a mountain pygmy-possum; can you confirm?'

Sure enough, looking up from inside the trap was the sweet face of a mountain pygmy-possum! There followed a scurry of sorting through the seed collection from Rough Creek, a presence confirmed and excited planning with Hayley and Haijing to commence trapping the northern sites.

The next two years saw us trapping every boulder field between Jagungal and the Bogong Peaks, 50 kilometres north of Cabramurra and parallel with the ACT. Easier now to pick them out on satellite imagery, which was not available when I started on the southern sites. Included was a huge boulder field on the flank of Snow Ridge, 800 metres long and supporting the largest colony of mountain pygmy-possums now known in NSW (134 individuals trapped in 2022) with Blue Cow-Guthega falling into second place at 70. Rough Creek also supported a sizeable population, with a smaller colony in Happy Jacks Valley concentrated along the lower slopes at 1,200m near the permanent river, cooled by cold air drainage.

We mounted an expedition to Boltons Hill in January 2011, repeated in October 2012. It was hot, alive with biting March flies by day and mosquitos by night. We bemoaned the death of the ancient snow gums on the summit, incinerated by a bushfire that passed through the area in 2003. There was little shade afforded to our tents by the scraggly regrowth from lignotubers of the few snow gums not killed.

Water was rationed and we had only one wet wipe each per day for ablutions. One evening the promise of a storm saw Hayley grab a bar of soap, strip and soap herself under the first few drops of rain which then ceased. In the following days there was great hilarity as every time she sweated, she bubbled!

The boulder fields were impressive, a few tall mountain gums flanking their edges having survived the fire. The boulders descended in broad sweeps of rock terraces, reinforcing the initial impression of rock waves. Possums were hard won and not plentiful, the conditions too hot and dry. Like in the valley below, few mountain plum-pines remained, most having been killed in the 2003 fires, leaving white skeletons stretching dead fingers over the rocks.

It was the 18th of January 2003 when I received the devastating news that Mt Blue Cow had burned. I watched the fire roar into the suburbs of Australia's capital, Canberra, causing tragic loss of life and property. Accessing Mt Blue Cow the following week, with smoke still rising from the dry, smouldering peat and ash shin deep, I could see around 50 percent of the mountain-plum pines had been burnt. Their branches like brittle bones still stretch across the rocks bleached white by 20 years of alpine weather. I had taken sections of trunk from the plum-pines killed at Mt Blue Cow and had them prepared and aged by our botanist and a dendrologist at the Australian National University. The oldest of those ancient, fire-sensitive shrubs, only 10cm in diameter, was 400 years. Twice the age of European arrivals to Australia and testament to the absence of fire from mountain pygmy-possum habitats in the past.

Haijing's research demonstrated the importance of the mountain plum-pine cover to cooling the rocks below, helping to make the rocks habitable for the possums (which cannot survive ambient temperatures above 28 degrees Celsius). Hayley showed that a permanent water source was the most important variable influencing the occupation of boulder fields by possums. Below around 1,000m the boulder fields are hot and dry and become engulfed in the soil and vegetation of the montane forests.

The Millennium Drought of 2001 to 2009 crept upon us gradually from its beginnings in 1997. In 1998, I published the population trends of the mountain pygmy-possums at Mt Blue Cow, concluding that there had been no discernible change in population numbers, despite some annual fluctuations. The wide-ranging movements of the animals had been maintained, although channeled, by constructing rock-filled travel corridors across the ski runs. However, at the end of 1998 the numbers of the broad-toothed rat, a gentle, fluffy and predator-naïve native rodent (Australia's answer to the guinea pig) plummeted, and they have not been trapped in the possum habitats since.

In 1999, the numbers of male possums on the monitored sites dropped from an average of 15 to three, followed in 2000 by a 50 percent decline of the females. I had encountered foxes and the occasional feral cat on my nocturnal forays around the mountain in 1987 and 1988, with cat encounters becoming quite regular by 1989 as the resort came into operation. In 1999, a litter of kittens was found in one of the main boulder fields. In 2002, with possum numbers still low, we began a winter cage trapping program for feral cats around the resort, catching 30 cats during the first winter.

4 January 2020

I am living on a property at Bywong, north of Canberra. The temperature has climbed to 43 degrees Celsius and the air is thick with smoke. I have the radio on and eyes glued to the fire mapping updates on my iPad. I have just heard the Dunns Road fire has developed into a catastrophic firestorm, sweeping through Cabramurra creating 128 kilometre per hour winds and temperatures approaching 70 degrees Celcius. The eye was centered on Rough Creek.

On January 15, with fires still burning towards Lake Eucumbene, I went into the area with a team of colleagues and NPWS staff. We took a truck load of feeding and water stations we had constructed while waiting anxiously for permission to enter.

We drove through kilometre after kilometre of blackened forest, snow gum woodland and alpine meadows, coming into Happy Jacks Valley to see all the regrowth from 2003 burnt again, but worse. The mountain gums on Bolton's Hill were now gone, but with some browned leaves still evident in patches of shrubs near the creek. Then to Rough Creek, where the fire had been so intense not even the fine branches of the snow gums remained and every single mountain plum-pine, having escaped the 2003 fire in the shelter of the rocks, was now destroyed. It was heart-breaking.

My cat monitoring cameras dripped plastic on the rocks, but remarkably one of the SD cards was intact, later showing the approaching flames then stopping abruptly. Another camera at Happy Jacks Valley showed a family of dingos, a kangaroo, a brushtail possum with a baby on her back, cats and foxes passing by on the road on their nightly business before the fire, and then the valley choked with fire and swirling ash. After the fire was an image of the juvenile possum looking distressed and then not seen again, the white mum dingo and her pup also missing, no kangaroo, but still a feral cat smiling smugly into the camera.

We worked in awe with purpose holding back dismay, installing the first of 61 feeding stations and 30 water stations that we deployed on the three sites in the hope that the pygmy-possums had survived the fire deep in the shelter of the rocks, and with the knowledge that they would have little to eat if they had. Due to the prevailing drought conditions, the streams at Rough Creek had ceased running and the bogong moths had failed to arrive from their breeding grounds on the parched western plains. The following week we included Snow Ridge, which had suffered a similar fate, with the camera there recording a temperature of 89 degrees Celsius before it stopped.

What finally brought the tears that day was stepping out of the vehicle on the way back past Lake Eucumbene, where the fire was steadily burning forward, and hearing a lyrebird singing in its path.

The conditions that led to that fire and Australia's Black Summer of 2019-20 were three years of extreme, unrelenting drought and high temperatures that hit us like a freight train, brought on by global heating. In the 37 years that I have lived and loved the mountain pygmy-possum and its alpine environment I have seen the snow cover become less reliable, shallower and melting earlier in spring. Mid-winter rain events often replace snow; water dripping into the hibernation sites of the possums wakes them and causes energy loss and death before spring. Fires are becoming a frequent event rather than a rarity. Drying streams result in less available habitat for the possums and food shortages are caused by drought, low numbers of bogong moths and loss of shrub cover, especially mountain plum-pines, from fires.

Feral cat numbers are increasing, probably due to warmer conditions in winter and fewer years of heavy snow fall that might kill them in hard winters. Bogong moth numbers, estimated to number around 4 billion in years past, plummeted over the last few years, leading to their listing by the International Union for Conservation of Nature in December 2021 as an endangered species. Many of the mountain pygmy-possums in the three colonies near Cabramurra miraculously survived the fire and we fed them with 12 kilograms of 'bogong biccies'—a biscuit developed by Zoos Victoria to match the nutritional value of the moths—weekly from January 15th 2020 until May 2022, with a break through the winters while they hibernated.

Due to our feeding efforts, these northern colonies have survived better over the last three years than the colonies in the unburnt south of the park, where food has been scarce due to low numbers of bogong moths. Fortunately, the drought broke in March 2020 and with three years of wet La Nina conditions, much of the surrounding shrubby vegetation has recovered and should provide enough food for the possums in following years. Bogong moth numbers were slow to recover, possibly because much of southeast Australia suffered severe flooding and their lowland breeding grounds may have been inundated, but numbers were increasing in spring of 2022. Sadly, the fire-sensitive mountain plum-pines were almost universally killed, and few seedlings are evident. We have a team of dedicated NPWS staff planting young plants propagated from cuttings, although given their slow growth rate it will take many years in the absence of fire before they are large enough to provide food and cover for the possums.

What is the future for the mountain pygmy-possum? Like many other wildlife species in alpine and lowland environments, it is facing severe threats from climate change with global heating and increased variability of climatic conditions. Climate change can affect wildlife through direct heating (e.g., flying foxes and Carnaby's black cockatoos falling dead out of trees during heat waves), loss of habitat, changes in the timing or abundance of their food sources (e.g., early flowering and fruiting of plants becoming mis-timed for migratory species) or increased impacts of native and introduced pathogens, diseases and insects.

Animals that can move to stay in their preferred climatic zone have some chance of dealing with a warming world. For example, some fish species are shifting from warm to cooler latitudes. However, species like the mountain pygmy-possum or the lemuroid possum of the wet tropics, are particularly vulnerable because they are already restricted to their climatic limit, isolated at the tops of mountains, and they have nowhere to go—a similar story to the distressing images of polar bears floating on melting ice floes in the Arctic, or penguins on collapsing ice shelves in Antarctica.

Threats from a warming world are very evident in the Snowy Mountains, where we have seen swathes of shrubs dying with the increasing activity of *Phytophthora* species (parasitic native soil fungi) due to the warming soils. Sadly, the few large tracts of the ancient snow gums which escaped the 2003 and 2020 wildfires are now dying due to an outbreak of the endemic longicorn beetles.

Climate change is a global threat and can only be addressed by a global response, but the response must start at home. This is particularly relevant to Australia, where in the last few years the word 'unprecedented' was used by some governments as an excuse for unpreparedness for both fire and flood. However, climate scientists have been predicting and warning about these conditions for the last 35 years or more.

There are less than 3,000 adult mountain pygmy-possums living in the small patches of boulder fields and surrounding shrubby heath sprinkled across the alpine areas of Australia's southeast—approximately 1,000 in New South Wales and the remainder in Victoria. The numbers on the monitoring sites at Mt Blue Cow fell to two females and one male in 2004 following the 2003 fire. In 2008, with continuing drought, only one female remained, high on the peak—and she had lost her tail from an unknown traumatic event. Following rains in 2010, the numbers increased steadily until 2017 when, with the shrubs (apart from the plum-pines) recovered from the fires and an ongoing program to trap feral cats and foxes, we trapped the highest number of possums ever recorded on the site (45 females and 18 males). We removed 202 feral cats from the resort between 2002-2022. We have dedicated teams of government agency staff, zoological institutions and volunteers helping to look after mountain pygmy-possums in both states. Our feeding efforts following the 2020 fires have demonstrated that in extreme events, we can keep them going for a short while. This gives me some hope that with our help this remarkable, resilient little possum will continue to grace our mountains in the future.

I had predicted in 1998 that global heating would see the possum gradually move up to the highest elevations with the rising snow line and the lower elevation populations would be the first to be lost. But ecology is never simple. What we are actually seeing is that the highest elevation populations are more susceptible to snow loss because of the extreme low temperatures without the insulating cover of snow, and scarcity of food if possums wake from hibernation. The possums at those sites are also more vulnerable to the loss of bogong moths during summer, as there are fewer alternative food sources at the high elevations.

The possums at the mid elevations have a greater chance of adapting due to alternative food sources and milder temperatures—if the streams keep running and we can keep the feral predators under control and an eye on the genetics. There is a captive breeding program in Victoria at Healesville Sanctuary and another starting at Secret Creek in the mountains west of Sydney, where we will attempt to breed them at warmer temperatures and test their resilience to a warming world.

The mountain pygmy-possum has received a great deal more attention than many endangered species, possibly because it emerged from the past quite recently. It wasn't until 1966 that it was first discovered by Europeans as a live animal in the Victorian alps. I'm sure the local First Nations people would have encountered them long before this during their seasonal forays to the mountains for ceremony and to gather bogong moths, a highly nutritious food. However, until its discovery at Mt Hotham it was known to Western science only as a fossil, a jawbone discovered in 1895 in a Wombeyan cave deposit by physician and palaeontologist Robert Broom.

It was given the name *Burramys parvus*, from the local Aboriginal burra burra, meaning 'place of many rocks'; mys and parvus are Greek for 'mouse' and Latin for 'small', respectively. A great many dedicated people are determined that this endearing, charismatic little possum, the only mammal in Australia confined to the alpine region and which hibernates throughout winter, is not once again relegated to the past. On a personal level it seems that my passion and involvement through much of my working career with 'Broom's little rock mouse' was a stroke of destiny.

It is my hope that every person in this heating planet of ours will realise the urgency of the climate crisis and the necessity to halt rising temperatures before it is too late for the mountain pygmy-possum, which has become an icon of the Australian Alps, and indeed for many creatures on this planet, ourselves included.

WHEN SMALL CHANGES ADD UP:
AN OBSESSION WITH THE RED HANDFISH

BY

JEMINA STUART-SMITH

Dr Jemina Stuart-Smith *is a marine biologist at the Institute for Marine and Antarctic Studies, University of Tasmania, Australia. She coordinates the Handfish Conservation Project and leads the National Handfish Recovery Team. Her work focuses on aiding the recovery of three critically endangered handfish species in Tasmania and restoration of their habitats. Jemina has spent several years diving around the world with the Reef Life Survey program and helped launch the national Redmap Australia marine citizen science program. She has a keen interest in threatened species research, marine conservation and restoration, and science education and engagement.*

We swim out along the surface, scanning the water below as we go. The strip of reef that runs alongside the rocky shore is relatively narrow and bordered by a deeper sand edge. We descend when we can see the sand, and reel out our transect line parallel to it, in about five metres depth.

We go along the patch of remnant seaweed clinging to the rocks; one of the last refuges for this little fish—the red handfish—and around this it is bare and bleak. Even if you had not heard stories, or knew of the past life of the reef, you would sense something was wrong if you dived there now. It's like walking into an unfurnished room in the middle of an otherwise occupied house. It almost has a sense of eeriness, an innate feeling that something is missing, even if you had not seen its previous state.

Even by the ocean's standards, handfish are peculiar looking little creatures. They possess two large fins resembling 'hands', a toad-like head that sports a grumpy-looking upturned mouth, and a punk-style dorsal fin reminiscent of a mohawk. But it doesn't end there. Their idiosyncrasies are coupled with a preference for walking on the seafloor using their over-sized 'hands' rather than swimming, and they have a fluffy 'lure' on their head, characteristic of anglerfishes, that they move around presumably as bait. Instead of a conventional gill—opening like other fish, they have a small pore behind their pectoral fins. Adding to these unusual behaviours, they are also ambush predators—preferring to sit and wait for food to swim past, rather than chase it themselves. These quirky features, combined with their elusive nature, make them more akin to an oddity from a children's fiction book, rather than a living creature.

My first dive at the red handfish site was in 2010, when I had only recently heard about this tiny, elusive fish through local divers and scientists in Hobart. The science and dive community are well-connected, something which often comes with living in a small island state. At the time, we only knew the red handfish from this single site. There were murmurs across the dive community regarding their small declining population and fears that they were headed for extinction.

I was mostly incognisant to their plight back then; keen to see them because they were so unusual and the rumoured urgency about running out of time. It didn't occur to me that no-one was working to protect them.

Human impacts like climate change are discussed at large scales by scientists, but small changes are often noticed first by local people, like divers on the east coast of Tasmania who have witnessed changes occurring in their lifetime.

At the time I had dived for a few years and was involved in a citizen science project called Reef Life Survey. It was as part of that program that I went out on my first dive to see, and help collect data on, the red handfish. I had already dived to help conduct surveys for another handfish species, the closely related spotted handfish. These fish also occur on the doorsteps of Hobart, albeit in totally different habitat-they are found in sandy, silty areas. They look and behave quite differently to the red but are just as beautiful.

The red handfish dive site is close to shore, just an hour from Hobart. The traverse to the water is down a short rocky slope that would be difficult to navigate at the best of times, let alone when laden with a thick wetsuit that restricts movement and carrying heavy dive gear and equipment. The shoreline is rocky, and when we reached the water, we sat resting our dive tanks on the rocks with our legs in the water to put our fins on and rinse our masks. The cool water was knee-deep as we slid in.

Even many years and dives later, at this point of every dive I think about the same thing. I picture the accounts of older divers who have told stories about seeing red handfish so numerous in these shallows that they numbered in the hundreds. In such times, I'm told, divers had to be careful of their footing when clambering in from shore for fear of standing on them in the shallows. The seaweed-covered reefs they lived on extended hundreds of metres along the shoreline; a vast area for these mysterious little creatures to hide.

Those stories of lush reefs supporting an abundance of red handfish occurred within my lifetime, but before I even knew handfish existed. Sliding baselines are gradual changes in the accepted norm of the environment, despite it being quite different from its initial state. It's where your original reference point, what you use to measure change, is in itself a degraded state. Each generation who views it then has an altered perception of what the condition is due to lack of personal experience in the original state. Changes you see are relative to your earliest recollection (of an already degraded state), rather than its true original state.

Today the reef no longer extends very far. The rocky structure that once bedded the plant life is still there of course, but it is largely devoid of the seaweed that used to cover it; the casualty of ecosystem imbalance at the hands of humans that resulted in an increase in native urchins (which consume the plants the fish need for cover), among other impacts that combine to degrade their habitat.

Our historical knowledge of other handfish species is also limited, although early European explorers and convicts collected and painted handfish—with some of these specimens remaining in museums across the globe. From this, scientists suspect they must have been quite common during the early days of British settlement in Australia. While I don't know if Indigenous cultures hold stories and information about handfish, I can't help but think it is very likely for those species that inhabit shallow waters.

I edge into the water, taking deep breaths, and accept the cold. The site sits in a little pocket that is influenced by currents and nearby water bodies; it gets cooler water temperatures in winter and stays warmer longer in summer, than most of the state. We think the cold winter temperatures are particularly important for the handfish.

We don't have enough data to understand exactly why the red handfish need the cold water. We theorise that it could be due to breeding or egg developmental cues, but we don't know enough about their biology, or where else they lived previously, to work this out. At this point, much of our effort is focused on conservation of the species as we tackle the looming threat of extinction.

Climate change threatens the persistence of many marine species. Some animals can adapt or adjust—move with warming waters or change the timing of their biological events (such as breeding or migration) to coincide with changing temperatures. But some species have limited ability to adapt. The red handfish walk on the sea floor; they lack a planktonic life stage which would allow them to follow currents to new areas, and both of these characteristics limit their ability to disperse in the face of climate change (or other impacts). They now occur in small, isolated populations—without the ability to reconnect. They also live on the south-eastern coast of Tasmania, at the bottom of Australia. With warming waters and no shallow reefs further south, there is nowhere for them to move to, even if they could.

More immediate, however, is the indirect threat caused by increases in the number of native urchins, which feed on the kelp reefs, reducing the habitat needed by the red handfish. The urchins are spreading because their predators—such as lobsters—have been removed. The red handfish are also close to urban areas and are likely impacted by pollution, siltation, and direct disturbance by humans.

I don't recall the first half of the survey on my initial dive there, the pre-handfish part, but I clearly remember getting halfway through my cryptic fish search, slowly sweeping away some thick brown seaweed to uncover a little red handfish sitting motionless in the centre of my field of view—staring up at me. I recall being startled. I had been so convinced that I wouldn't find one that I then froze when I saw it.

Neither of us moved, a trait I'd later learn is just the norm for the red handfish. I didn't have a camera, so I just stayed still, staring, trying to make a mental note of every tiny detail of this little creature. The bright red fins contrasting against the dark shell-grit on which it was perched, the fluffy lure on its head, and that characteristic frown on its little face. And the size! Small enough to fit in the palm of your hand, smaller than the spotted handfish I'd seen previously. I stayed for as long as I could.

My first encounter was captivating and exciting, but it wasn't until years later that I was given the opportunity to work more closely with red handfish.

While I continued doing volunteer citizen science, it wasn't always around the red handfish. In fact, I probably didn't revisit them again for several years, focusing instead on diving abroad and on mainland Australia more so than in my home state. But several years later I fell into a role with the National Handfish Recovery Team that arose from concern about the declines of the species (from the monitoring we had been doing). For reasons I'm not entirely sure of, I felt a sense of responsibility to be involved in the recovery of this species.

This new role meant that several years after my first dive at the red handfish site I began returning there regularly. The dives are long, because the handfish are difficult to find, and we move slowly and carefully as we search to minimise disturbance caused to them. We sometimes spend more than 100 minutes underwater. We do a mix of handfish-specific dives for a PhD project and continue the Reef Life Survey transects intermittently.

A typical Reef Life Survey transect involves laying out a survey reel (a measuring tape), swimming along counting fish, then focusing search efforts on a narrow band on the sea floor to look for all the things that live there or like to hide. The searching is slow and meticulous. It requires carefully looking down crevices and brushing aside seaweed to find the sea creatures that don't wish to be found, for which the red handfish is the poster child.

The divers who volunteer are often those who like to focus on the little things: the nudibranch-seekers, those who spend half a dive taking photos of the intricate patterns found on a seastar or finding millimetres-long amphipods clinging to seaweed.

For them, it's all about the hunt; finding that new or unusual species, behaviour, or colouration. They often lose themselves in the moment so much that during the cryptic fish search, seals and sharks might swim unnoticed overhead. It makes them the ultimate handfish detectors. I wouldn't technically put myself in that category, but I do like the little things—which might be why I became so fascinated with the handfish in the first place.

Over time, the handfish have become harder to find, and our search area even smaller as the seaweed patch containing the fish continues to diminish. But then, just a few years ago, there was a community sighting of a red handfish in a new location, and a second population was discovered by the Reef Life Survey team, also close to Hobart.

This second site holds more handfish, and with it, hope for the future. It has allowed us to start a conservation program to increase the wild population. We have trialed collecting eggs to hatch and raise young in captivity—to protect them while they grow, before returning them to the wild. We take photos of spots on their sides and have given some of them tiny tattoos before we release them, so we can identify them when we see them again.

There has only been hatch and release so far, but the work continues. The new site is buying us time while we look at other options, including maintaining an 'insurance' population in captivity. We've also taken steps to reduce urchin numbers, to help restore the handfish habitat. It's temporary fix, but it's another way to buy us a little more time.

Our sites are often littered with discarded fishing gear-sinkers and lines that we remove. The lobsters that once must have been abundant are all small. Boats whir past, and the odd diver stops by to photograph handfish or collect a feed of shellfish. The human impacts on the natural world are always visible, but fewer people see these consequences since they're underwater. Climate change presents an ever-looming threat to the red handfish, as the cold water they thrive in is reduced to small pockets. But human-induced changes to the ecosystem, and other human disturbances, exacerbate the danger to this already critically endangered fish.

On a good day, with three hours underwater, we might be lucky enough to find two red handfish at our original site. I have been diving there intermittently for 10 years now and that is all I have ever known—my sliding baseline. Yet despite the low numbers, the cold conditions and often limited visibility, it is never tedious. Each sighting is met with excitement and avid descriptions as we dismantle our dive gear—where it was, what it was doing, whether it was one we'd seen previously, or curious behaviours that we might not have noticed previously. Perhaps it is the rarity that makes every sighting exciting, and because we fear we are looking at a disappearing species. Whatever the reason is, every find represents a glimmer of hope, when on some days we're not sure we will see anything.

DRIVING AT NIGHT WITHOUT HEADLIGHTS:

CLIMATE CHANGE IMPACTS ON THE WEST AFRICAN CANARY CURRENT

BY

TODD CAPSON

*Dr **Todd Capson** has 27 years combined experience in South Africa, West Africa and Latin America building and managing programs to address the impacts of climate change on the oceans, to strengthen marine protected areas and to build scientific capacity. As a Research Associate of the Paris Institute of Global Physics, his work addresses the impacts of climate change on West African oceans. He is now part of a team working to establish South Africa's first marine UNESCO World Heritage Site. As an AAAS Diplomacy Fellow at the US Department of State, he served on the Interagency Working Group on Ocean Acidification and led their international shark conservation program. At the Smithsonian Tropical Research Institute, he led efforts to protect Panama's Coiba National Park through national legislation and its recognition as a World Heritage Site. He also led an international team to discover new medicines from tropical ecosystems. He supervised three master's-level students as an Adjunct Professor of McGill University. He has authored or co-authored six book chapters and 42 peer-reviewed articles in the fields of climate change, organic chemistry, biochemistry, the discovery of novel medicines in tropical ecosystems and chemical ecology.*

I n March of 2022, during a US Embassy-sponsored trip to Senegal, I visited the fishing town of Joal, located roughly 100 kilometres south of Dakar. Joal is Senegal's largest fishing port and is linked by a bridge to Fadiouth, an island constructed entirely of discarded clam shells. The vast majority of the combined population of Joal and Fadiouth, around 50,000 people, live off the traditional fishing industry. The beach of Joal is lined with colourful *pirogues*, open fishing boats that are the vessel of choice for Senegalese fishermen, and which feature in many photographs of the country's coast.

In Joal, I met with Mariam Sy, a gracious, no-nonsense woman who runs a vast operation that preserves fish by smoking, drying, salting or fermenting. Fishing, and the processing and sale of those fish, is the main source of income in Joal. In Senegal, men do the fishing while women are responsible for processing. Mariam lamented that her husband has to work much harder to catch fewer fish than in the past and that she now has to import fish from elsewhere in Senegal and from Guinea-Bissau in order to sustain her cooperative, greatly increasing her costs and reducing profitability.

There is no question that overfishing plays a role in decreasing catches in Senegal, but largely absent from the conversation, and on every level—scientific, among non-governmental organisations (NGOs) and in policy circles—is the recognition of the role of climate change. That carbon dioxide (CO_2) emissions can have profound impacts on marine life, especially in upwelling ecosystems like the one bordering northwest Africa, is irrefutable. But without data on how CO_2 emissions are changing the biogeochemistry of the oceans, Mariam and others that depend upon the marine resources of the region are effectively driving a car at night with no headlights, with no idea of what is in front of them or how to respond.

I marvelled at the disparity between the hard-working people of Joal—who don't appreciate the threat posed by CO_2 emissions on their livelihoods and food security—and the situation in the US Pacific Northwest, with comparable oceanographic conditions but where the coastal waters are bristling with state-of-the-art instruments operated by PhD-level scientists from government and academia, providing the data on changing ocean biogeochemistry that identify threats to fisheries and aquaculture that are used to inform management decisions and adaptation efforts.

I promised Mariam that I would work to draw attention to the plight of Senegalese fisheries and to find resources to study the role of climate change. For example, if the government had data that made clear the impacts of climate change on fisheries, it could reduce other pressures on fish stocks, such as removing access to Senegalese waters by foreign fishing fleets from Europe and Asia. But without rigorous data on the impacts of climate change on marine resources, the problem is 'out of sight, out of mind'.

The oceans have absorbed roughly 30 percent of the CO_2 produced from the burning of fossil fuels, resulting in what Nicolas Gruber of ETH in Switzerland calls the 'triple whammy': rising temperatures, ocean acidification and ocean deoxygenation. These stressors act synergistically, impacting molecular processes, organisms and ecosystems, in ways that we are only beginning to fathom, and that are irreversible on the scale of human societies.

Oceans are warming by the same greenhouse effect that we see on land—a blanket of CO_2 produced by the burning of fossil fuels traps the heat just like a blanket traps your body heat while in bed. A warming atmosphere heats the ocean surface which interferes with ocean circulation and keeps oxygen from the atmosphere from entering the ocean depths. As water warms, it also holds less oxygen. While ocean warming, acidification and deoxygenation are globally acting processes, these phenomena are far more pronounced in some regions, making them 'hotspots' for climate change impacts on the oceans.

Among those hotspots are the planet's four Eastern Boundary Upwelling Ecosystems (EBUEs), located along the eastern basins of the Pacific and Atlantic Oceans, bordering the west coasts of North and South America and Western Africa. EBUEs are among the world's most productive ocean ecosystems, a consequence of the upwelling of cold and nutrient-rich waters, providing around 20 percent of global fish catches; but the waters are also high in CO_2 and oxygen-depleted. When the carbon dioxide from the burning of fossil fuels is added to the waters of an already CO_2-rich EBUE, biological thresholds can be exceeded, and with profound consequences as described below in the California Current System, an EBUE that extends from southern British Columbia, Canada, to Baja California, Mexico, including the Pacific Northwest.

While working on the US Interagency Working Group on Ocean Acidification, I learned of the collapse of the Oregon-based Whiskey Creek Shellfish Hatchery, that once produced 10 billion larvae of the Pacific oyster, *Crassostrea giga*s, for the West Coast shellfish industry. Starting in 2007, the hatchery began to experience larval mortalities of up to 80 percent. One of the hatchery owners, Mark Weigart, described how the inlet pipes to his hatchery, which had to be routinely cleared of shellfish and barnacles in the past, are now clean save for a few misshapen barnacles. It is now well documented that the accounts I heard from Mark and others are due to ocean acidification. Among the consequences of acidification is a decrease in pH but also in the pool of carbonate, a component of calcium carbonate, a fundamental building block for the skeletons and shells of shellfish and corals, plus countless other organisms that play key roles in marine ecosystems, livelihoods and food security. A key article by Richard Feely and colleagues in the journal Science documented for the first time the arrival, to the surface off of northern California in 2006, of waters that are corrosive to calcium carbonate, corroborating that the mortality of oyster larvae in the Pacific Northwest was due to ocean acidification.

While the mortality of the Pacific oyster may be due in part to the fact that it is not native to the California Current System, the same cannot be said for species that are native to the region. Large portions of the shelf waters are now corrosive to the shells of the sea butterfly (*Limacina helicina*), a delicate sea snail just five millimetres across which underpins key marine food webs that sustain herring, salmon, whales, seals, seabirds and other species. Severe carapace dissolution and destabilised mechanoreceptors were also observed in the larval Dungeness crab (*Metacarcinus magister*) which supports a fishery that generates annual revenues up to $220 million.

Also, in Oregon in 2006, anoxic waters upwelled to depths of less than 50 metres within two kilometres of shore, where they persisted for four months, and have returned every year since. There were no prior records of such severe oxygen depletion over the continental shelf in the region. Starfish and mussels died, and rockfish and other mobile fish fled the hypoxic zone, which grew to 3,000 square kilometres, as Virginia Gewin described in the aptly titled feature in *Nature,* 'Dead in the Water'. Bear in mind as well that the CO_2-rich, oxygen-depleted waters that upwelled off the coast of the Pacific Northwest and were responsible for the events summarised above, were last exposed to the atmosphere some 60 years earlier, when atmospheric CO_2 was downwelled into the oceans. This is a grim reminder that ocean acidification and deoxygenation will be manifest long after humanity stops burning fossil fuels.

In learning about the impacts of CO_2 emissions on the California Current System, I would come to appreciate the crucial role of monitoring the biogeochemical manifestations of climate change: increasing acidification, deoxygenation and temperature. Monitoring these phenomena in such dynamic upwelling systems must be 24/7 and *in situ*; namely, with instruments in the water where the phenomena are manifest. It was because of such a system that scientists were able to document the upwelling of corrosive, anoxic waters in the California Current System, and how stakeholders like Mark were able to understand, anticipate, and to some degree, adapt to the impacts of climate change on the region. Through the rigorous and continuous monitoring of the water used to grow oyster larvae at the Whiskey Creek Shellfish Hatchery, and by buffering that water to obtain conditions that allow the larvae to survive, the hatchery has been able to recover much of its original levels of production.

Mark's gut-wrenching accounts of the impacts of ocean acidification on his shellfish hatchery drove home to me the point that humanity has actually managed to change the chemistry of the oceans, tapping into a deep emotional attachment I have for the sea that is shared with many others, and is why I work to raise awareness about climate change through its impacts on marine life and ecosystems. The power of that message was evident during a workshop I organised between shellfish growers and scientists from the US Pacific Northwest and New Zealand. My overall goal in organising the event was to help spread the word that climate change impacts on the oceans are profound and happening now.

The highlight of the workshop was when Alan Barton, who works with Mark at the Whiskey Creek Shellfish Hatchery, shared his first-hand observations about the impacts of ocean acidification on their hatchery. Alan is plain-spoken and smart: he was among the first to point the finger at ocean acidification when oyster larvae were dying, and other potential sources of mortality were ruled out.

Alan told the audience, 'The only way we're going to stop acidification is to get out in front on this issue—nobody else is going to do it, and we're the guys who will first go out of business. So, I encourage you to get on board with this issue, convince yourself first and get out there and convince other people before it's too late.'

He drove the point home with a cartoon showing school kids hiding under their desks while mushroom clouds were seen out the schoolhouse window. A pin dropped on the floor of the workshop venue would have landed with a thud after Alan's talk. That the message was delivered by someone from the aquaculture industry as opposed to a banner-unfurling environmental activist (who indeed have key roles to play) removed any doubt from any erstwhile climate change deniers attending the workshop.

Also impactful was a presentation by an oceanographer from Oregon State University, George Waldbusser, who discussed the importance of the first 48-hour window that follows fertilisation of an oyster egg: baby oysters have two days to make a shell—without it they dissolve. While there are variables at play, such as the amount of wind-induced upwelling of CO_2-rich waters, in a nutshell, the two-day window for the fertilised eggs of the Pacific oyster to make a shell in the untreated waters of the California Current no longer exists.

I set out to learn whether the phenomena I learned about in Oregon were happening elsewhere. The Canary Current Large Marine Ecosystem (CCLME), which extends along the northwest African coast from the northern Atlantic coast of Morocco to Guinea-Bissau, is another EBUE and hotspot for climate change impacts. The CCLME has complex geography and ocean circulation and is the least studied or understood of the four EBUEs.

From 2016 to 2019, I lived in Senegal in order to better understand the impacts of climate change on the Canary Current, the level of awareness of the threats posed by climate change impacts on the oceans and to learn about any research programs dedicated to understanding the issue. There are roughly 63 million residents of countries bordering the CCLME who are highly dependent upon marine resources for food security, national economies and livelihoods. In Senegal, fisheries provide 75 percent of animal protein for people and employ one-fifth of the working population. The same animal protein helps sustain the most vulnerable communities in sub-Saharan Africa, nearly 200 million people. West African agriculture is highly vulnerable to climate change due to its reliance upon rain, the region's climate variability and limited economic and institutional capacity to respond to climate change. This will likely lead to an even greater dependence on fish for food security and livelihoods in the region, a dependence that will be exacerbated by population growth.

As a consequence, the countries bordering the West African Canary Current are 'between a rock and a hard place', caught between climate change impacts on land and sea, where social and climatic factors intersect to create an acute climate vulnerability. Senegal is one of the most stable countries in Africa, with three peaceful political transitions since independence in 1960, but as a lower-middle-income country, poverty is ubiquitous. From the people I met through work, like Mariam from Joal, and from personal acquaintances, like the friends I played music within the bars of Dakar, I was constantly reminded that many Senegalese live on a knife-edge. On some days they enjoy a Senegalese classic, *Ceebu Jen*, a rice and fish dish, and on other days, there isn't enough to eat. And many are far less fortunate.

But unlike the situation in the California Current System, where rigorous data informs scientists, policy makers and stakeholders, in the CCLME region (with the exception of a single mooring managed by French researchers) there are no permanent assets to monitor the impacts of climate change. The absence of data means that there is also no awareness of the potential threats that rising acidification, deoxygenation and warming pose to fisheries. Exacerbating the problem for West African fisheries is overfishing, both legal and illegal; weak monitoring, control and surveillance; offshore oil and gas extraction; land-based pollutants; and a competing global demand for fishmeal that is likely to grow.

Observations such as the first report of a nearshore anoxic event on the west African shelf in 2012 (off the coast of Senegal) and millions of dead fish washed ashore in Mauritania in 2020, which scientists attributed to a lack of sufficient oxygen, provide strong evidence that the CCLME is indeed a hotspot for climate change. In the face of multiple threats, the effective management of fisheries in the CCLME will require a thorough understanding of the impacts of climate change, including rigorous, in situ data on increasing acidification, deoxygenation and warming. For example, if a fishery or shellfish stock collapses, stakeholders won't know whether it was a consequence of overharvesting, deoxygenation that cause fish to migrate to more oxygen-rich waters, acidification-induced shellfish mortality, or some combination of these or other factors.

While living in Dakar, and working in partnership with French and Senegalese scientists, I put into place a two-prong strategy to address the issue of climate change impacts on the oceans of the CCLME. First, we have worked to draw attention to the issue by writing editorials in local newspapers and giving lectures to officials in government and academia. Second, we have looked for funding for the training of African scientists and the purchase of instruments. It is likely to take years before this strategy yield results. But I have learned that there are always funds for a good project—it's a question of persistence and finding the right donor—and I am more persistent than the tide.

Scientists from Chile have developed a research program that could serve as a model for understanding and adapting to climate change impacts on the Canary Current. Chile borders the Humboldt Current System, an EBUE that extends along the west coast of South America. Chilean researchers monitored ocean acidification along the length of the Humboldt Current and have studied its impacts on local shellfish species. They have discovered shellfish strains that are relatively tolerant to ocean acidification together with optimal habitats for their cultivation, providing a potential means of cultivating shellfish in future, and likely more acidic, oceans.

Shellfish have been harvested in Senegal for at least 5,000 years. In some Senegalese coastal communities, shellfish harvesting and processing are the main source of income for vulnerable groups, and shellfish are among the most accessible sources of animal protein for some populations. The island of Niodior is located south of Joal and is part of a region known as the Sine Saloum, home of extensive mangrove forests. The exploitation of shellfish on Niodior has enabled the women to generate a minimum of 14 percent and up to 100 percent of their total household incomes. In addition, oyster farming has been identified by the government of Senegal and the World Bank as a key sector for sustainable development growth in Casamance, the southernmost region of Senegal.

Like the Humboldt Current System, the Canary Current System has a broad diversity of habitats and oceanographic influences that could reveal strains that are relatively resistant to acidification and suitable for cultivation. Development of shellfish aquaculture in countries such as Senegal could draw upon the wealth of experience in the California Current.

The Californian Current System and Humboldt Current System show how data on ocean acidification and ocean deoxygenation can identify threats to economically important fisheries and aquaculture, inform adaptation efforts and identify relatively acidification-resistant shellfish species and optimal habitats for their aquaculture. Adaptation to climate change impacts on either land or sea for the countries bordering the CCLME will be difficult and expensive, but *anticipatory,* data-driven adaptation would be much more cost effective than reactive adaptation. The data generated from in situ monitoring should be of 'climate quality'—suitable for the detection of long-term anthropogenically driven changes in ocean chemistry, a requirement for many international climate assessments and for international data repositories. Data of this quality would provide an accurate picture of climate change impacts on the CCLME, facilitate comparisons between those impacts and other regions on the planet and allow policy makers and resource managers to make informed and defensible management decisions.

Climate models are widely used to project future changes in ocean health and are included in policy documents such as the UN Intergovernmental Panel on Climate Change (IPCC) *Special Report on Climate Change and the Cryosphere in a Changing Climate* (SROCC), which provides 'the best available scientific knowledge to empower governments and communities to take action…' The SROCC draws upon the scientific literature, including model studies that compare how different regions of the world's oceans will respond to climate change.

But for models to be useful predictive tools, two conditions must be met. First, they must be backed by solid in situ data in order to validate the predictions of that model. Second, the data used to validate those models must be on a scale that that takes into account the oceanographic variability of the region. In the CCLME, neither of those conditions are met, meaning that their projections for the region are incomplete at best and misleading at worst. Nevertheless, there are model studies published in prestigious journals such as Nature Climate Change that predict positive or minor impacts of climate change for most or all of the countries bordering the CCLME. An overarching problem is that model studies on a global scale that include regions such as the CCLME can give the false impression that the scientific community can provide meaningful projections on the impacts of climate change on fisheries and marine ecosystems. I am once again drawn to the staggering disconnect between people like Mariam Sy and, in this case, scientists in wealthy countries who develop sophisticated computer models on state-of-the-art computers, but whose projections are fundamentally disconnected from the nitty-gritty reality of Joal.

An essential step to building the capacity required to effectively anticipate and adapt to changing ocean chemistry will be the training of additional African PhD-level scientists. Local scientists are best positioned to ensure that the data obtained will inform policy and practices on local and national levels, and provide the basis for long-term, effective and proactive responses to climate change. While international capacity-building entails commitments of time and money that go beyond the research programs typical of scientists from wealthy countries, there are models of north-south partnerships among institutions of higher education that include capacity building in the developing world. Examples include those between West African institutions and the French National Research Institute for Sustainable Development, the German government's West African Science Service Centre on Climate Change and Adapted Land Use program, and the Norwegian government's Ecosystem Approach to Fisheries (EAF)-Nansen Project.

Wealthy nations rely upon the data from biogeochemical monitoring programs to develop models and policies that provide guidance to industries and local stakeholders. The West African countries bordering the Canary Current, for whom climate change impacts on the oceans will impact livelihoods, food security and development outcomes, deserve no less. Greater awareness about climate change impacts on the oceans would also add more voices to the global chorus calling for reductions in CO_2 emissions.

The time I spent with Mariam and her colleagues left me with a profound sense of purpose. It is simply unacceptable to watch their livelihoods, and the ecosystems they depend upon, slip away while more privileged scientists and policy makers refuse to recognise the reality of their circumstances. And there are proactive ways to respond to a changing ocean, as described above.

It reminded me of a time in Panama, some 20 years earlier, when I sat on the beach of an impossibly beautiful tropical island at the centre of the Coiba National Park. The president of Panama had just issued an executive decree that would have resulted in the destruction of the park's marine and terrestrial ecosystems. I was a scientist at the Smithsonian Tropical Research Institute and my portfolio didn't include saving tropical ecosystems. I screamed into the wind, 'Where the fuck are the NGOs that get paid to save places like this?'

But I couldn't find an NGO willing to take on the president, so I did it myself. The Coiba National Park is now protected by national legislation and as a UNESCO World Heritage Site, and widely recognised in Panama as a jewel of biodiversity.

Looking at the Senegalese coastline from the window of my airliner as I departed in March of 2022, I thought to myself, 'Here we go again.'

CLIMATE CHANGE GIVES NEW MEANING TO 'GLACIAL PACE'

BY

MARTIN TRUFFER

Professor Martin Truffer *grew up in the Swiss mountains. He studied Physics in Switzerland, where he discovered that studying glaciers could be an actual job. He then moved to Alaska to pursue a PhD in Geophysics, investigating how glaciers move over their substrates. After a short stint with the Australian Antarctic Division, he managed to persuade the University of Alaska to hire him as a Professor in Physics, with a research focus on glaciers and ice sheets. His research has taken him to remote field sites from Greenland to Antarctica and many of the glaciated places in between. He focuses on glacier dynamics, in particular phenomena associated with fast glacier flow and the interaction of ice flow with the ocean.*

The clouds were slowly lifting as we ascended the mountains behind my home town in Switzerland on an early morning hike. My teenage daughter and I joined my sister to look for her goats and bring them some salt to lick. As we ascended the mountain slope, we soon emerged from the forest and were rewarded with views of our local glacier. I don't come here often, but each time I do, I look in astonishment at the additional land that has been exposed by the rapidly retreating ice. A bit nostalgically, I pointed out the area where we used to do glacier and ice training with our Alpine Club, back when I was my daughter's age. There is not a sign of glacier ice left at that location and the idea of even going to that particular area of steep and unstable rock seemed totally absurd.

The irony of the situation was not lost on me. I can well recall my Dad telling me similar stories about glacier crossings in places where it simply seemed unthinkable by the time I was a teenager. The glacier, and its growing and waning, play a big part in the history of our town. As we hike back down, we pass a chapel that was built in 1672 by my ancestors to plead with God to protect them from the advancing ice. As part of the deal, on the first Sunday in September each year, a procession is organised that ends in the celebration of the Holy Mass at this chapel. The town promised to abstain from cursing, rage, jealousy and disagreements, and to distribute the nourishing glacial water in peace. In addition, they were to make a donation of 'well-baked' bread to the poor and at least one member of each family was to participate in the procession and mass, which at that time went to the glacier and back. The procession still happens annually, and I have fond memories of participating, probably mostly because of the treats that were served and the whole town being out on an often magnificent early fall Sunday.

Most people look at this old custom with amusement and it is hard to believe now how scared people must have been of the advancing glacier. During the Little Ice Age, it was extremely tough to carve out a living in this harsh landscape. Old reports talk of years without summers. The land could not provide for everybody, and many had to leave and find a living elsewhere. The glacial melt water was crucial to the livelihood of this agricultural society in the driest part of Switzerland. The oldest preserved town documents pertain to strictly regulated contracts about the use of this water, and the maintenance of the many waterways that form a complicated network to deliver it from the glacial creek to the fields. Even during my time, one family member was busy for much of the summer with watering the fields.

When the advancing glacier threatened those waterways, it was time to ask for divine intervention. This worked much better than anybody could have anticipated, and many people think that by now we ought to be praying for the glaciers to grow again. In fact, a nearby town with a similar oath has asked the Pope to officially reverse it and instead pray for an end to further warming.

When I was hiking in the local mountains as a kid, I would never have thought that one day I would have a job studying glaciers. I was fascinated by the moving ice, the noise it made, and the raw power of nature it represented. During my studies, I realised that mathematics and physics provided me with the tools to investigate how glaciers work. It was the early 1990s and at that time I had little scientific interest in climate change; it seemed like a distraction from what was really interesting about how glaciers behave.

My studies eventually brought me to my second home in Alaska, where I completed my PhD and then found a job that has brought me to many glaciated places all over the globe. During this time, I have had the opportunity to witness changes on a much bigger scale than those I was familiar with at home. For example, in the early 2000s, we received funding from the US National Science Foundation to study the calving of ice into a lake near the town of Yakutat, Alaska. The 337 square kilometre Yakutat Glacier was thinning by several meters per year. It had been shedding large icebergs and retreating into the 69 square kilometre Harlequin Lake.

We were interested in finding out why so many lake-calving glaciers in Alaska were retreating so rapidly. Within the three years of the project beginning, more than three square kilometres of additional glacier ice simply disappeared, as it broke up into big icebergs that were then flushed into the ocean. Our research object simply melted away.

As we continued our studies, we concluded that the entire ice field feeding this glacier is doomed, and—even under no additional warming—only small remnants will be left by the end of this century, giving way to an entirely new landscape and a big lake. The reason for this unfavourable scenario is that the entire glacier is very deep and is resting on bedrock or sediment that is very near sea level. So, as the glacier thins, its mean surface elevation decreases, which exposes it to increasingly higher temperatures. That surface drop might not appear like much in the beginning, but over a decade or more it adds up. This creates a feedback loop that is detrimental to the glacier: the more it thins, the warmer it is on its surface, which leads to even more melting.

At Yakutat Glacier, the thinning has been so large that it has reached a point of no return. That is, no additional warming is necessary for the glacier to disappear entirely, and with each additional year of thinning, the amount of cooling it would take to reverse course keeps getting larger. A 'normal' glacier has the option to escape this feedback by retreating back up a mountain slope to avoid the high temperature and melt of the lower elevations and to 'try' to adjust to a new equilibrium. But some of these large deep glaciers may not have that luxury.

The story of melting and disappearing glaciers will sound familiar to many by now, even if the extent of the change is often difficult to picture. We do know that glaciers are melting. We hear it in the news, we read it in scientific papers in which the mass loss is carefully tallied and converted into how much it will raise sea levels. We measure the extent of ice with unprecedented accuracy using ever-better technology from space, from airplanes and on the ground. Very precise lasers are mapping out the Earth's surface at regular intervals and numerous satellites are used to make maps and compare them to each other so that we can inventory the health of thousands of glaciers worldwide.

But even as a scientist who works on and thinks about glaciers for every day of my professional life, it is always astounding to see how these numbers actually play out in the landscape. I have been lucky enough to be able to study glaciers in many of the world's icy areas, including the big ice sheets in Greenland and Antarctica. As a young student I pleaded my way into joining an expedition to Greenland's largest glacier, Sermeq Kujalleq, or Jakobshavn Isbrae as it is better known. I remember well flying into our camp and looking out from the big helicopter at the enormous mass of ice making its way towards the ocean. We were here to continue a program to find out what enabled this glacier to move so fast, more than 10 meters per day near the front. Our task was to use hot water to drill holes into the ice and put in instruments to measure the deformation and temperature of the ice. I was super excited to be part of such a project and at the prospect of spending five weeks on the ice.

I had never seen anything like it: ice as far as the eye could see in every direction. And yet, there was always something interesting to break that seemingly monotonous icescape. Cold rivers of meltwater were meandering down towards a large lake or disappearing in enormous gaping holes, called moulins, that lead right down to the bottom of the ice. An occasional bird would find its way to camp, perhaps blown off its course by a storm? As the ice was moving over the ground, it followed some of the wallops and bumps in the bedrock, so that the surface actually undulated much more than I initially perceived. A half hour walk from camp could get one to a spot of complete solitude in the middle of this vast ice sheet, a feeling that keeps bringing me back to such places.

Our work there was successful. We drilled several holes, some of which reached all the way to the bottom of the ice, and we left instruments frozen into the ice. A year later my friend and a team returned to retrieve the data, which formed a major part of his PhD thesis. He showed how the fast flow in this more than 2,500-metre-deep ice-filled fjord was caused not only by the deep warm ice, but also by the presence of much more deformable dirty ice that was a result of dust being blown onto the ice sheet during the last ice age. The warm ice at depth is a result of the heating the glacier experiences as it deforms rapidly, much like a churning dough warms up.

This was all very exciting stuff, but I was still thinking of this ice sheet as a huge slow giant with some fast moving parts, nothing different that I would see change in my lifetime. That belief had certainly been influenced by what I had been taught in my university classes and read in textbooks: yes, glaciers are changing, but the big ice sheets have much longer reaction times and will take centuries to show any real change. Boy, was I wrong!

A decade later my colleagues and I secured our own funding to go back to Greenland. Things had changed in a major way. The ice sheet didn't agree with my thinking of it as a slowly changing behemoth. Sermeq Kujalleq was unrecognisable upon our return. A 15-kilometre-long stretch of ice was no longer there. It had broken up into huge icebergs and floated out to sea. The ice that was lost had worked like a plug on the glacier draining this huge ice sheet. With the plug removed, the ice was now moving at three or four times its former speed and rapidly drawing down the ice sheet. We were camped on the side of this massive fjord and measured survey markers on the ice moving at 40 meters per day, putting a new meaning to 'glacial pace'.

We witnessed enormous pieces of ice breaking off and slowly rolling over. The first time we saw it I kept clicking away on the camera and it wasn't until later that we realised the full scale of what had just happened: a junk of ice, almost a kilometre deep, more than half a kilometre long and a few hundred meters wide had just broken off the glacier. Never before, I think, had such an event been caught on camera. The following year we brought lots of cameras and set them up to record pictures every 10 seconds to record several such events. The energy released in such big calving events is enormous, equivalent to a small atomic bomb. With our measurements we helped solve the mystery of how these glaciers create slowly evolving seismic signals that can be detected globally. The sleeping giant of the Greenland Ice Sheet was awake. No self-respecting glaciologist would ever think of ice sheets as slowly evolving again.

And things were not any different on the other side of the globe. Some glaciologists had warned us as much as 50 years ago that vast parts of Antarctica are barely stable and that several meters of sea level rise are possible if that ice becomes unstable. A particularly vulnerable section of Antarctica had been called the 'weak underbelly'. This vulnerability was recognised because of the very deep base of the ice in this sector of Antarctica, which is known as the Amundsen Sea Embayment. Ice penetrating radar had revealed that sections of this ice sheet were grounded at well over 1,000 metres *below* sea level. Some researchers had already identified an instability that is now known as the Marine Ice Sheet Instability: as the area where ice first meets water—the grounding zone—retreats into a deeper basin, it will start flowing more rapidly. This additional flow of ice leads to thinning and therefore ungrounding of more ice, and off we go.

This mechanism can become unstoppable until the ice retreats far enough so that the bottom gets shallower again. In the 1990s, satellite imagery started to show that the Amundsen

Sea Embayment was indeed an area of rapid change. But it remained largely unexplored. The logistical difficulties of getting to this very remote place, notorious for bad weather, seemed unsurmountable. But evidence kept piling up that the Amundsen Sea is the one point from which the largest amount of ice could be dumped into the ocean over a relatively short time period. Furthermore, careful tallies pointed to missing mass: the amount of ice breaking into the ocean was much less than that lost from the ice sheet. The mystery was solved when it became clear that the ocean is warm enough to melt large amounts of ice, where it first comes in contact with the ocean and from the underside at the floating extension of the glacier.

This had to be investigated further and I had the opportunity to travel to the two largest glaciers in the area: the Pine Island Glacier (somewhat affectionately known as the PIG) and, more recently, the Thwaites Glacier, one of the planet's biggest glaciers. We were part of a large international collaboration to investigate this glacier, because it is generally recognised as being one of the prime potential contributors to sea level in the next decades to centuries.

After a long journey that went via New Zealand and the US Antarctic station of McMurdo, we finally landed a plane on the Thwaites Glacier and established a field camp that became our little home for the next few weeks. We were at a location on the lower glacier where the ice was about 400 metres thick and floating on the ocean. Nothing on the surface indicated that we were so close to sea water.

Once again, I found myself at a location of seemingly all-encompassing whiteness. But despite the immense solitude and quiet, things were moving along rapidly. I made a habit of taking a waypoint on my handheld GPS every night after I had crawled into my tent, noting each time that we had moved another two metres or so towards the ocean. But because everything around us was moving at the same pace, one would never really have guessed.

Our purpose on the ice was to use hot water to drill holes all the way to the water below. It was a strange experience; after several hours of watching a drill hose disappear into a hole, we made contact with the underlying ocean. We quickly proceeded with measuring ocean properties, such as salinity and temperature, that would tell us how the ocean eats away at the ice from underneath. We sent a camera and a light down the dark abyss, all the way to the ocean floor, where we took sediment cores that will hopefully reveal the recent glacial history of this area. The camera was even visited by a little squid who expressed its displeasure by spraying ink all over it! We had to work quickly, because the ice was quite cold and the holes would freeze in just a few hours. Before that happened, we left a string of instruments, connected it to a fancy weather station and configured it to send data back to the US via satellites, so they could be analysed later in the comfort of our offices.

These data will help us understand where the water that flows under the ice shelf comes from. We will be able to refine weather and tide models for the area, which are currently largely unconstrained by data. We can watch the changes of water temperature and salinity that are so important for the health of the ice. But for now, we still had to pack up all our gear and prepare for the long journey home. Standing on the flat white ice in occasionally fierce snowstorms, it was hard to fathom that the viability of coastal towns all over Earth is tied to what this piece of ice is going to look like in the next few decades. We were standing in one of the most remote places on Earth and yet what will happen there will determine the fate and livelihood of millions of people.

I never sought out climate change in my research. I simply wanted to learn more about glaciers, how these wondrous things work, what makes them speed up and slow down, and how they create the mountain landscape that I love so much. But it feels like climate change sought me out. It follows me everywhere I go. It saddens me when I see the projections of near total ice loss in the European Alps by the end of the century. It might be silly to mourn the loss of a glacier, but there are dire consequences, both to local water resources and natural hazards, not to mention global sea level. And I would still take some comfort if my kids don't have to explain to the next generation how they used to hike up to this glacier—and then have to explain what a glacier actually is.

AFTERWORD

— BY —

TERRA ROAM

Terra has been on the road for more than 30 years. Their travels through work, volunteering and record-breaking expeditions have taken them around the world, working closely with a wide variety of cultures, and deep into the wilderness alone studying nature.

Wow! That was a wild ride of adventure on the high seas of climate change.

As I volunteered in a Victorian BlazeAid camp during flooding, Wendy asked if I would write an afterword for this anthology. To be honest, I was expecting academic science reports, which I don't mind, so it was a great surprise to find myself completely enthralled by the fascinating stories each expert shares with us.

It was also a good distraction from the seriousness of volunteer work helping farmers recover after another huge flooding event.

Earlier this year, I helped out at a Blaze Aid camp in Casino, Northern Rivers of New South Wales, after their devastating floods. Each morning, in our work teams, we would travel out and meet the farmers for a chat. These morning meetings were not only to receive instructions, we also used them as a farmer well-being check. I hadn't been in Victoria long before I noticed the difference between farmers who had already experienced major flooding events and those in Northern Rivers who were experiencing their first.

Farmers know the increase in floods, fires and drought is caused by climate change; their lives and livelihoods revolve around weather and seasons, and they are intrinsically connected to their environment. They also know they have very little control over the effects of climate change; we waited too long to implement effective change and the bulk of responsibility lies with our governments and big business. This criminal lack of action has resulted in increased natural disasters such as fires, floods, drought and cyclones, which are now threatening lives, food security and the economy. I have never met a farmer who denies the role of climate change in the disasters they are experiencing.

On one of my previous treks across the Australian Alps, I read *Three Men in a Boat (To Say Nothing of the Dog)* by Jerome K. Jerome, published in 1889, and *Far from the Madding Crowd* by Thomas Hardy, published in 1874. I was surprised that both authors reference climate change's visible impact on farming and rural life. Jerome writes about the impact of the Industrial Revolution on the countryside as his protagonists float down the Thames and Hardy describes seasonal changes and hotter summers in Wessex. I began reading further back and was not so surprised to see observations as early as 1780. We knew what was happening to farmers 240 years ago.

Like farmers, I have an intuitive sense of change in Australia's wild alpine environments. It is an intimate familiarity as I allow myself to become an animal, letting go of our human superiority and becoming a natural, curious and vulnerable part of the ecosystem. Full prolonged sensory immersion, sometimes for months at a time.

For the last 34 years, since moving to Kosciuszko National Park in the winter of 1988, the alps have had a magnetic pull on my soul. I was only 16 and spent every spare minute of daylight exploring the mountains, shoving a pocketknife, juice box and Vegemite sandwich in my pockets, sometimes taking a backpack with a jumper and rain jacket. I'd head off into the hills to follow animal tracks and creeks, swim under waterfalls, visit my tree friends and talk to wombats. This was long before the age of PLBs and mobile phones, but so long as I returned for dinner, nobody cared how far I wandered—in fact, it was encouraged. By the time I finished school in 1990, I knew the mountains from Lake Jindabyne to Mt Kosciuszko, between the Snowy and Thredbo Rivers, in all seasons like the back of my hand.

This brings me to something which weighs heavy on my heart and causes a great deal of anxiety. Almost every year, I return for one to three months, usually in late summer but sometimes to camp a winter season on the snow in my hammock or tent. Time spent in the mountains has included four crossings of the Australian Alpine Walking Track in four seasons, three times solo. The most recent leading Expedition Climb8, snowshoeing and trekking 610 kilometres from Wee Jasper to the north of Brindabella National Park to Barry Saddle in the Victorian Alpine National Park. Expedition Climb8 was my response to something many of us are feeling in these years of uncertainty. Since 1988, I have witnessed undeniable changes in climate patterns, snowfall, the snowline, rainfall, lightning strike fires, wildlife behaviour, feral pestilence and vulnerable alpine habitats above the treeline.

For the last two decades, since noticing the change in climate patterns, I have been prompting action from those with most at stake, as well as those who share in the responsibilities of cause and response. The lack of action—and outright denial—from the highest stakeholders only gave me more anxiety. I feared for the future of Australia's precious alpine regions and humanity's stupidity.

In 2004, one spring break while studying conservation ecology, I hitchhiked from university into the mountains. I walked over the top from Dead Horse Gap to Rainbow Lake, off-track along the ridge, looking for signs of Ngarigo, Ngunnawal and Yuin summer camps and paths. This was a personal interest that I hoped to incorporate into a few subjects and, in the process, learn as much as possible about the First Nations people of the northern alpine regions.

I crossed through the mountains and made my way to the National Parks office in Jindabyne. I fronted up and explained I was researching Aboriginal use of the Australian Alps. They let me hang around for a week, which I mostly spent buried in the archives, researching the shamefully small amount of information they had. I began looking into climate change, parks and resort policies, and historical evidence from colonisers, farmers, gold miners, meteorologists and botanists. While I had access to the archives, I used them to my advantage. It did nothing to allay my fears; in fact, it made it worse.

This was how I met Dr Phil Zylstra—author of the first

chapter of this anthology—at the national parks headquarters; he was lumped with babysitting the 'work experience kid'. Phil wasn't a Dr yet, he was a fire officer studying fire behaviour and soon went on to create his behaviour model for Australia's many varied ecosystems, weather and climate zones. It was Phil who taught me to find solutions to problems rather than flogging methods which don't work and putting up with the same results.

Dr Phil taught me not to put up with what can be changed. I have made it a life motto: 'Action is the best antidote to anxiety.'

Since then, I have embarked on many campaign walks for various causes, from kelp forests, renewable energy, suicide prevention and most recently Climb8, more than 25,000 kilometres of campaign treks. A few years of research goes into each walk, and a few decades of research went into the expedition across the mountains. It was a turning point, learning about many scientists and the results of their findings. Dr Linda Broome's exciting news about Burramys parvus, the mountain pygmy possum, adapting to a warmer climate at lower altitudes gave me a real sense of hope. As did a day on country with an Aboriginal teacher and scientist, talking about fire, land management, the importance of our connection to the land and listening to the spirits of the old ones.

In the end, climate change drastically changed my plans for Expedition Climb8. The tragic fires of summer 2019/20 closed Namadgi and large sections of Kosciuszko and the Victorian Alpine National Parks, most of the Australian Alpine Walking Track had been affected and more than half was closed. These closures were both to give the fragile ecosystems a chance to recover without humans stomping through and to keep humans safe from tree fall. I had to design a new path following back roads, snowy hydro service roads and forestry tracks, through the Brindabella mountain range and across private farms to avoid the high-risk sections to the north.

In the south, where I continued on alone, I waited until the tracks were cleared and re-opened. It was the worst snowfall on record—there was so little snow some of my team for the northern section didn't even get to use their new snowshoes. Although this was a minor issue compared to what the small marsupials reliant on snow cover for insulation and protection from predators were dealing with. Spring plant and animal behaviour began alarmingly early, and places normally covered in deep snow were being ploughed up by feral pigs and horses. The tiny seedlings and epicormic growth emerging after the fire had a new battle.

We had planned to summit as many peaks as possible during Climb8, but the snow cover was too thin and my team were constantly falling through and tangling in the tundra. This was despite the pandemic lockdown pushing the start of my expedition back into July when there should have been more snow.

Although the expedition experienced unprecedented conditions, I used the platform as a conduit for positive change. I opened up discussions about anxiety, adaptation and action, not only looking at natural sciences but the economy, community and recreation connected to alpine climate change. It helped me manage my anxiety, which was beginning to spiral, and it helped others along the way, giving people hope and something else for their mental health toolbox.

As the intrepid scientists within these pages have shared, we are looking at changes that are significantly impacting us. What we do with this research shapes our future and the health of our beautiful planet Earth. It is easy to get lost in the big picture, but don't let go of all the epic work and solutions being found by the pioneering and dedicated researchers. From the depths of the oceans to the mountaintops and beyond to satellite imagery and monitoring, we have learnt so much. Thinking of how much more we will discover as technology advances the future is exciting, and more solutions are waiting for us to find.

INDEX

Thank you to Paddy Pallin for sponsoring this book.

At Paddy Pallin, concern for the environment is not just an important issue; it's a way of life, a commitment that governs many aspects of the company. It's an inseparable part of the ethos of Paddy Pallin, and one of the company's core values.

Through lending support to this anthology, Paddy Pallin continues this commitment by helping to educate and inspire action. Together we recognise the importance of empowering people with knowledge as we rally to put the planet first.

Paddy Pallin

SINCE 1930